The Molecular Biology of the Gene
(1965, 1970, 1976, coauthor: 1987, 2003)

*The Double Helix: A Personal Account of the
Discovery of the Structure of DNA* (1968)

*The DNA Story: A Documentary History of
Gene Cloning* (coauthor: 1981)

Recombinant DNA (coauthor: 1983, 1992, 2007)

The Molecular Biology of the Cell
(coauthor: 1983, 1989, 1994)

*A Passion for DNA: Genes, Genomes,
and Society* (2000)

*Genes, Girls, and Gamow: After the
Double Helix* (2002)

DNA: The Secret of Life (2003)

AVOID BORING PEOPLE

AVOID BORING PEOPLE

LESSONS
FROM A LIFE
IN SCIENCE

James D. Watson

ALFRED A. KNOPF
NEW YORK
2007

THIS IS A BORZOI BOOK
PUBLISHED BY ALFRED A. KNOPF

www.aaknopf.com

Knopf, Borzoi Books, and the colophon are registered trademarks of
Random House, Inc.

Library of Congress Cataloging-in-Publication Data

Watson, James D., 1928–
Avoid boring people: and other lessons from a life
in science / by James D. Watson.—1st ed.
p. cm.
ISBN 978-0-375-41284-4 (alk. paper)
1. Watson, James D.—Biography.   2. Molecular
biologists—United States—Biography.   3. Scientists—
United States—Biography.   I. Title.
QH3.W34.A3 2007
572.8092—dc22
[B]   2007015675

Manufactured in the United States of America
First Edition

For Paul Doty

Contents

Foreword

Robert Maynard Hutchins once famously remarked that whenever he felt the urge to exercise, he immediately lay down. Despite the hyperbole (for he, like Jim Watson, was an ardent tennis player) he had made the point that his priority resided in one relentless form of exercise, that of the mind. In this, too, Jim Watson is a true follower of Hutchins, as his book and indeed his life demonstrate so engagingly.

Under the leadership of Hutchins, the University of Chicago played a significant role not only in Jim's education, but as a permanent benchmark in his continuing thinking about education. If one of the principal goals of a liberal education is to leave students with an insistent lifelong preoccupation with the question of what education is for and what it is about, what qualities should characterize the truly educated person and what intellectual virtues should be sought and practiced—and why—then Jim is a triumphant example of its success. Hutchins was the great critic of and innovator in twentieth-century higher education. He made his university the home of perhaps the most passionate debates over curriculum and intellectual purpose to take place in recent memory. And Hutchins, like Jim, was never satisfied: his university, he said, might not be very good, but it was the best there was. For undergraduates, "the best" had to do with a vision of general education that introduced students to the most profound questions of human life and civilization, and to the great writers and thinkers who had confronted and tried to make sense of them—an education that taught students to think with rigor and integrity and to go on doing so under whatever circumstances they might encounter.

The budding ornithologist who entered Chicago's college left as a man ready to embark on the study of the gene, focused on an ambitious specialization while always underlining the need to sustain the widest possible intellectual curiosity. The imperative that Jim Watson had learned to embrace was to pursue the most difficult, perhaps intractable, problems and to think scrupulously and express his views honestly, whatever might be the consequences in terms of what is now called "political correctness," other kinds of mere conformity and self-interest, or even *politesse*.

The austerity suggested by such an outlook does not, fortunately, preclude a fine sense for the enjoyment of the world, its pleasures and its foibles. Jim portrays himself as a very young and rather monastic undergraduate and as something of a late, and utterly enthusiastic, social bloomer. The freshness and candor of his observations on people and events, accomplishments and follies (including his own) may well derive from the novelty he found in every experience that, new to him, made life ever more interesting.

This combination of qualities shone through *The Double Helix* as it does here, and indeed Jim's account of the struggle to get the manuscript published more or less intact is almost as entertaining as the book itself. And just as *The Double Helix* became a classic in showing how science may really get done, so the present volume, in describing the whole trajectory of a unique career within a larger world of science and research, will join it and add to our understanding of Jim Watson's revolutionary achievements.

The different sides of Jim Watson's persona emerge vividly also from the "remembered lessons" detailed at the end of each chapter. Some may prove more useful than others. The advice to "Expect to put on weight after Stockholm" is not, alas, relevant for most readers, but "Work on Sundays" does have larger application. So, too, the wisdom of "Don't back schemes that demand miracles."

The maxim "Don't use autobiography to justify past actions or motivations" defines the captivating tone of this autobiography and its direct self-revelation quite wonderfully. At the same time, the exhortation "Avoid boring people" is scarcely necessary, for if Jim Watson is incapable of anything, it is of boring people.

In consequence, we must all be grateful to Jim for so exuberantly following his own lesson: "Be the first to tell a good story."

—*Hanna H. Gray, University of Chicago*

Preface

Most of my unpublished writerly output—handwritten manuscripts and letters, lectures and lab notes, university and government documents that I helped prepare—is deposited in the archives of Harvard University and of Cold Spring Harbor Laboratory. In time, it will all become available to the public, mostly through the Internet, for those preternaturally curious about how I have moved through life. Rather than let other commentators have the first crack at those writings, I opted to be the first to employ them extensively to prepare this look at my life before middle age became obvious—my childhood, university years, career as an active scientist and professor, and my first years as director of the Cold Spring Harbor Laboratory.

As this book's broad features came to cohere, I began to see *Avoid Boring People* as an object lesson, if not quite an exemplary history of the making of a scientist. It is my advice in the form of recollections of manners I deployed to navigate the worlds of science and of academia. The thought that this instructive value might be made explicit in the form of self-help led me to conclude each chapter with a set of "remembered lessons"—rules of conduct that in retrospect figured decisively in turning so many of my childhood dreams into reality. Suffice it to say, this is a book for those on their way up, as well as for those on the top who do not want their leadership years to be an assemblage of opportunities gone astray.

Skipping high school's last years led to my never learning how to type, and even today I generate left-hand-written versions of the first drafts of all my writings. Without my administrative assistant Maureen Berejka's ever-increasing skill in handling strings of seemingly indecipherable squiggles, this book could never have come into existence. In preparing successive early drafts of the manuscript, I much benefited from the able Barnard College chemistry graduate, Kiryn Haslinger, whose expert knowledge of the English language led to many improvements in my word use. New York University psychology graduate Marisa Macari ably provided help in inserting period photographs and documents. Later, Stanford biology major Agnieszka Milczarek invaluably corrected the many errors of fact and spelling spotted by friends to whom I sent preliminary drafts. Finally, I much thank George Andreou of Knopf for masterly editing that has much improved this volume's clarity and intellectual thrust.

—Jim Watson, March 26, 2007

AVOID BORING PEOPLE

1.   MANNERS ACQUIRED
   AS A CHILD

With my mother in 1929

I **WAS BORN** in 1928 in Chicago into a family that believed in books, birds, and the Democratic Party. I was the firstborn, followed two years later by my sister, Betty. My birth was at St. Luke's Hospital, not far by car from Hyde Park, where my parents lived after their marriage in 1925. Soon after Betty was born, my parents moved to South Shore, a middle-class neighborhood populated by bungalows, vacant lots, and two-story apartment houses. We lived in an apartment on Merrill Avenue before relocating in 1933 to a small, four-room bungalow that my parents bought at 7922 Luella Avenue, two blocks away. That move allowed my by then financially stretched, seventy-two-year-old grandmother to live with us in the rear bedroom next to the kitchen. In newly created tiny attic rooms, Betty and I slept on bunk-like beds.

Though I went to prekindergarten at the Laboratory School of the University of Chicago, the Depression quickly placed private education beyond my parents' means. I, however, was in no way disadvantaged by changing to public school. Our Luella Avenue home was only five blocks away from the academically rigorous Horace Mann Grammar School, which I attended from the ages of five through thirteen. It was then a relatively new brick building built in the early 1920s in Tudor style, possessing a large auditorium for assemblies and a gymnasium where I was seldom able to do more than two or three push-ups.

While both of my father's parents were Episcopalians, only his father, Thomas Tolman Watson (born 1876), a stockbroker, was a Republican. His wife, constantly upset at being a speculator's wife, showed her displeasure by invariably voting for a Democrat. She was born Nellie Dewey Ford in Lake Geneva, Wisconsin. Her mother was a descendant of the settler Thomas Dewey, who arrived in Boston in 1633. The Watson side of my family can be traced back to the New Jersey–born William Weldon Watson (born 1794), who would become minister of the first Baptist church established west of the Appalachians, in Nashville, Tennessee. When he returned from a Baptist convention at Philadelphia in a prairie schooner (before railroads), he brought with him the first soda water fountain ever seen in Tennessee. To counteract the local whisky demon, he set up the soda fountain on a street corner near the church and single-handedly made soda water all the rage. Reputedly he made enough money selling soda water to build a new church for his growing congregation, and it still stands today in the heart of Nashville.

His eldest son, William Weldon Watson II, moved north to Springfield, Illinois, where it is said he designed a house for Abraham Lincoln across the street from his own. With his wife and brother Ben, he later accompanied Lincoln on the inaugural train to Washington. Ben's son, William Weldon Watson III (born 1847), married in 1871 Augusta Crafts Tolman, the daughter of an Episcopalian banker from St. Charles, Illinois. Afterward he became a hotelkeeper, first north of Chicago and then in Lake Geneva, Wisconsin, where he raised five sons including my grandfather, Thomas Tolman Watson. After my grandfather's marriage in 1895, he initially sought his fortune at the newly discovered Mesabi Range, the great iron-ore-bearing region located near Duluth on western Lake Superior. Then he joined his older brother, William, later to be one of Mesabi's senior management. My father, James D. Watson Sr., was born in 1897, followed over the next decade by his brothers, William Weldon IV, Thomas Tolman II, and Stanley Ford.

From northern Minnesota, my father's parents moved back to the Chicago region, where, with the help of his wife's money, my grand-

My father, James D. Watson, Sr.,
in 1925, the year he married my mother

My mother, Margaret Jean Mitchell
Watson, models the MacKinnon
kilt from Scotland.

father bought a large Colonial Revival–style home in Chicago's afflu-
ent western suburb of La Grange. My father went to the local schools
there before attending Oberlin College in Ohio for a year. Dad's freshman year ended, however, with scarlet fever rather than academic lau-
rels. The following year he would start commuting into Chicago's
commercial banking center (the Loop, with its ring of elevated tracks)
to work at the Harris Trust Company. Money was short, as usual.

But making money was never near my father's heart, and after World
War I started, he enthusiastically joined the Illinois National Guard
(Thirty-third Division), soon shipping off to France for more than a
year. Upon coming home he began working at La Salle Extension University, a prosperous correspondence school that offered business
courses. There he met, in 1920, his future wife, Margaret Jean Mitchell
(born 1899). She came to work in the personnel department after fin-

ishing two years at the University of Chicago. Mother was the only child of Lauchlin Alexander Mitchell, a Scottish-born tailor, and Elizabeth (Lizzie) Gleason, the daughter of an Irish immigrant couple (Michael Gleason and Mary Curtin) who had emigrated from Tipperary during the potato famine of the late 1840s. After farming ten years in Ohio, they moved on to land south of Michigan City, Indiana, and it was there, in 1860, that my grandmother, Nana to Betty and me, was born.

Early in her adolescence, Nana had left the Gleason farm to become a servant in the Barker mansion, the home of Michigan City's most prosperous family, owners of a large boxcar factory. Soon Nana rose to be Mrs. Barker's personal maid, accompanying her around the spas of the Midwest. Later, Mr. Barker took closer note and installed her in a Chicago flat together with funds that allowed her an independent life. Only more than a decade later, in her mid-thirties, did she marry Lauchlin Alexander Mitchell, born in Glasgow (1855) to Robert Mitchell and Flora MacKinnon.

As a youth, Lauchlin Mitchell immigrated to Toronto and from there to Chicago, where his custom-made-suit business thrived by the time of the Columbian Exposition of 1893. Sadly, he died in an accident when my mother was just fourteen, struck by a runaway horse carriage while coming out of the Palmer House Hotel on New Year's Eve. My mother's only mementos of him were a tiny MacKinnon kilt sent to her from Scotland and the splendid pastel of him that was exhibited at the 1893 World's Fair, reportedly commissioned in exchange for a custom-made suit. Nana began to take in paying guests, in effect running an Irish boardinghouse on Chicago's South Side.

Marring my mother's adolescence was a lengthy bout of rheumatic fever that damaged her heart and made her short of breath if she exercised seriously. The illness ultimately cut her life short and she died of a heart attack at the age of fifty-seven. Like her mother, she followed the Catholic religion but was never a serious churchgoer. I remember her going to mass only on Christmas Eve and Easter, always saying that her heart needed to rest on weekends. On many days, particularly Sunday, Nana helped prepare our family's meals, having a skill at cooking not commonly found among the Chicago Irish. Her presence in our home early on gave Mother the freedom to work part time at

Elizabeth Mitchell in Michigan City
in 1925, before she was my nana

the Housing Office of the University of Chicago, helping to supplement my father's barely adequate salary, which had been cut by half to $3,000 when the Depression took hold.

Behind our house was an alley that separated the homes on the west side of Luella Avenue from those on the east side of Paxton Avenue. The general absence of cars made it a safe place for games of kick-the-can or setting off firecrackers that could still be bought freely around the Fourth of July. When I finally began to grow past five feet, a backboard with a basketball hoop was put up above our garage doors, allowing me to practice my free throws after school. Scarce family funds also purchased a ping-pong table to liven up winter days.

My irreligious father only reluctantly had agreed to a Catholic baptism for my sister and me. It kept peace between him and my grandmother. He may have regretted the accommodation when my sister and I began going to church on Sundays with Nana. At first I didn't mind memorizing the catechism or going to the priest to confess my venial sins. But by the age of ten, I was aware of the Spanish Civil War and my father let me know that the Catholic Church was on the side of the fascism he despised. Though one priest at Our Lady of Peace gave sermons supporting the New Deal, many in the congregation bought

Horace Mann Grammar School kindergarten class photo, October 1933.
I am seated on the floor, second from the left, proudly wearing a bow tie.

Father Coughlin's vitriolic magazine opposing Roosevelt, England, and the Jews, which was on sale outside Our Lady of Peace following each Sunday mass.

Just after my confirmation at age eleven, I completely stopped going to Sunday masses in order to accompany my father on his Sunday morning bird walks. Even as a small boy I was fascinated by birds, and when I was only seven my uncle Tom and aunt Etta gave me a children's book on bird migration, *Traveling with Birds*, by Rudyerd Boulton, curator of birds at Chicago's Field Museum of Natural History. My father's devotion to bird-watching went back to his high school days in La Grange. It continued equally strong after World War I when his family moved to an apartment in Hyde Park on Chicago's South Side, so his brother Bill could conveniently attend the University of Chicago. Dad would be up before sunrise most spring and fall days to go birding in nearby Jackson Park. Our first bird walks were also in Jackson Park. There I first learned to spot the common winter-resident ducks, including the goldeneye, the old squaw, the bufflehead, and the American merganser. In the spring I quickly learned to differ-

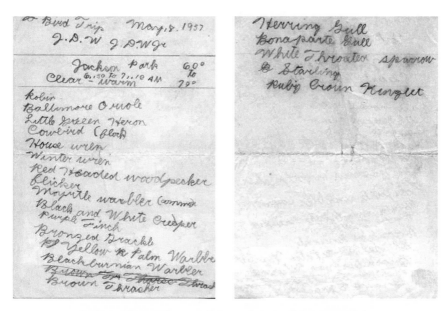

Early on I learned from my father to keep careful notes on birds.

entiate the most common of warblers, vireos, and flycatchers that in springtime migrated north from their tropical winter homes. By the age of eleven, I already had enough book knowledge of birds to anticipate the many new species we would encounter during a 1939 drive in a borrowed car to see the San Francisco International Exposition. On that trip I added more than fifty new species to my acquaintance.

My mother, by then a captain of our Seventh Ward precinct, enjoyed working for the Democratic Party. Our basement became the local polling station, earning us ten dollars per election, and my mother made another ten manning the polls. At the 1940 Democratic National Convention, held in Chicago, we rooted, to no avail, for Paul McNutt, Indiana's handsome governor then bidding to be chosen as Roosevelt's running mate.

In the evenings, Dad often was consumed with work brought home from the office. His principal task as collection manager of the LaSalle correspondence school was to write letters dunning students for delinquent payments. He never believed in threatening them, instead cajoling them with reminders of how their studies of the law or accounting

Ten years old and soon to abandon Sunday mass for Sunday morning bird-watching

would help them advance to high-paying jobs. I now realize how difficult it must have been to keep the job, he being a socialist-leaning Democrat sympathetic to the students who couldn't pay. No one, however, could accuse him of not working hard or of undermining free enterprise or, for that matter, of frowning upon the plutocratic game of golf, which he first came to enjoy in youth but later could play only on company outings after the Depression had forced the sale of the family Hudson.

Our family always rooted for Franklin Roosevelt and his New Deal's promise to rescue the downtrodden from the heartless grasp of unregulated capitalism. We were naturally on the side of the strikers in their violent confrontations with U.S. Steel at the massive South Side mill two miles east of our home along the shore of Lake Michigan. Matters of economics, however, began to concern our family less as the German menace grew. My father was a strong supporter of the English and French, the Allies on whose side he had fought in the First World War. He would have found the Germans a natural enemy even without Hitler.

I remember his anguish when Madrid fell to Franco. The local radio stations played up the defeat of the communist-backed republicans, but Father then saw fascism and Nazism as the real evils. By the time of Munich, the news from Europe was as much cause to be glued to the radio as was the Lone Ranger or the Chicago Cubs. Particularly crucial to us was the outcome of the 1940 presidential election, wherein Roosevelt sought his third term and was opposed by Wendell Willkie. Seemingly almost as awful as the Nazis themselves were America's isolationists, who wanted to stay out of the problems of Europe. My father was among those who saw in Charles Lindbergh's visit to Germany a manifestation of anti-Semitism.

Occasionally my parents had nasty disagreements about misspending what little money they had. But these frustrations were never passed on to my sister and me, and we each regularly received five cents for the Saturday afternoon double features at the nearby Avalon Theatre. Our parents would join us for some movies, one such occasion being John Ford's adaptation of Steinbeck's epic chronicle *The Grapes of Wrath*. The message that great decency by itself does not generate a happy ending never left me. A long drought that turned fertile farmlands into dust clouds should not cause a family to lose everything. How any responsible citizen could see this film and not see the good brought about by the New Deal boggled my imagination.

I always liked going to grammar school and twice skipped a half grade, graduating when I was just thirteen. Somewhat downcasting were the results of my two IQ tests, discovered by stealthy looks at teachers' desktops. In neither test did I rise much above 120. I got more encouragement from my reading comprehension scores, which placed me at the top of my class. I graduated from grammar school in June 1941, just after Germany invaded Russia. By then Churchill had joined Roosevelt as a hero of mine, and on most evenings we listened to Edward R. Murrow reporting from London on the CBS news. That summer was broken by my first time away from my family, going by train for two weeks in August to Owasippe Scout Reservation in Michigan above Muskegon on the White River. There I enjoyed working for nature-oriented merit badges that made me a Life Scout. Less fun were the overnight camping trips, in which I would invariably lag

Russell Hart (center) and I on the way to Boy Scout camp in 1941

behind the other hikers, catching up only when they stopped to rest. Nonetheless, I came home content to have spotted thirty-seven different bird species.

I could not help realizing, however, that as a boy expert on birds I was far behind the much younger Gerard Darrow, who thanks to a prodigious memory became a Chicago celebrity when as a four-year-old ornithologist his story and talent were written up in the *Chicago Daily News*. I would resent him even more when he became the first famous "Quiz Kid" of a Sunday afternoon radio program that first aired in June 1940. Groups of five children, each of whom received a $100 defense bond, were asked questions by the host, the third-grade-educated Jolly Joe Kelley. Previously he had been emcee of the *National Barn Dance,* having first come to radio fame by reading the funnies on WLS. Almost instantaneously, *Quiz Kids* became a national sensation, with its weekly listenership between ten million and twenty million—almost half the giant audiences that listened to Jack Benny, Bob Hope, and Red Skelton.

Virtually every Sunday afternoon for over two years, I listened to *Quiz Kids* hoping that somehow I might get on the program and win a

As a Quiz Kid in 1942, second from the left, doomed
by my lack of familiarity with the Bible and Shakespeare

war bond. Stoking this hope was the fact that one of the show's producers, Ed Simmons, lived in the apartment house next to our bungalow. Finally, because of a successful audition or because of Ed Simmons's influence, I became a fourteen-year-old Quiz Kid in the fall of 1942. My first two appearances went well, with lots of questions suited to my expertise. But my third time on, I was up against an eight-year-old called Ruth Duskin and a battery of questions on the Bible and Shakespeare dominating the thirty-minute horror. I had never been encouraged to know the plots of Shakespeare, and my early Catholic upbringing had furthermore shielded me from any knowledge of the Old Testament. So it was virtually preordained that I would not be one of the three contestants to come back for the next program. When we went home I bitterly felt the want of encyclopedic knowledge and quick wits needed for semipermanence as a Quiz Kid. But I was nevertheless three defense bonds richer. Later these were used to purchase a pair of 7×50 Bausch and Lomb binoculars to replace the ancient pair my father had used as a birder in his youth.

By then I was a sophomore at the newly built South Shore High

*Dad and I spotted the seldom-seen white-winged
scoter off Jackson Park in Lake Michigan.*

School. I continued as a largely S (superior) student, though I encountered much more competition than I had at Horace Mann. There was a great Latin teacher, Miss Kinney, who sent me off to the state Latin exams with the more stellar student Marilyn Weintraub, on whom I had a slight crush that no one ever knew about. I was then acutely conscious of my size, only five feet tall when entering high school, and shorter than my sister, who went through puberty early and reached her final height of five feet three inches while I was still only five foot one.

I worked on and off behind a neighborhood drugstore soda fountain making Cokes from syrup and carbonated water. The other conventional teenage job possibility was as a bicycle newspaper delivery boy. But that would have prevented my going on early bird walks with my father and was never seriously considered. Particularly in May, we would routinely get up while it was still dark so we could arrive in Jackson Park soon after sunrise. That way we would have almost two hours to go after the rarer warblers, principally in the region of Wooded Island. Dad's ear for birdsongs was much better than mine—upon hearing, say, the rough sound of the scarlet tanager, he never mistook it for the more melodious Baltimore oriole, which also migrates into Chicago just after the leaves come out. Afterward, Dad would catch a northbound streetcar to his work, while I would get a trolley going in the opposite direction, which would let me off near school.

It was in Jackson Park in 1919 that Dad had met the extraordinarily talented but socially awkward sixteen-year-old University of Chicago student Nathan Leopold, who was equally obsessive about spotting rare birds. In June 1923, Leopold's wealthy father financed a birding expedition so Nathan and my dad could go to the jack pine barrens above Flint, Michigan, in search of the Kirkland warbler. In their pursuit of this rarest of all warblers, they were accompanied by their fellow Chicago ornithologists George Porter Lewis and Sidney Stein, as well as by Nathan's boyhood friend Richard Loeb, whose family helped form the growing Sears, Roebuck store empire.

I had just entered high school when Dad and Mother first told me of Nathan and how for a thrill he and Richard Loeb had brutally killed a younger acquaintance, Bobby Franks. After offering Bobby a ride home from school, they fatally struck him in the head, disposing of the body in a culvert adjacent to the Eggers Wood birding site to which Dad later often brought me. Some six months before Nathan's senseless though nearly perfect crime of May 24, 1924, his father contacted mine expressing worry about his son's obsessive fascination with Loeb. Nathan by then was already at the University of Chicago Law School and Dad had no acquaintance with the rich college youths that Leopold and Loeb moved among. In July, Leopold and Loeb were defended in a Chicago courtroom filled with newspapermen by the celebrated Clarence Darrow, who asked Dad to appear as a character witness. But his family advised against this action, saying it would mark him for life in Chicago.

After Darrow had saved his clients' necks—they were sentenced to life imprisonment without possibility of parole—Nathan wrote Dad proposing a correspondence. But Dad never replied, still horrified by the crime that had gripped Chicago as none before. Many Leopolds and Loebs changed their names, and my father and Sidney Stein ceased all communication. Many years later, I came across Stein searching, as I was, for May warblers and flycatchers in the dunes near Waukegan above Chicago's North Shore. By then a very successful investment banker, later to be a trustee of the University of Chicago, Stein was acutely embarrassed when I identified myself as Jim Watson's son.

There was constant talk at home about the University of Chicago, especially since my father knew its president, Robert Hutchins, whose own father had been a professor of divinity at Oberlin when my father was an undergraduate there. Hutchins had recently enacted an exciting plan for admitting students who had only finished two years of high school and whose brains had not already been rotted out by the banality of high school life. Mother took the lead in seeing that I took the scholarship exam, administered one winter morning in 1943. Soon after, I was invited back to the campus for a personal interview, at which I talked about the books I had lately read, concentrating on Carlo Levi's antifascist statement, *Christ Stopped at Eboli*. Afterward I was very nervous until the director of admissions, a friend of my mother's, reassured her that I had a decent chance at a full-tuition scholarship. When I got the good news officially, I was too happy to care that my good fortune might have been related to my mother's being well liked by the members of the scholarship committee. Moving on to a world where I might succeed using my head—not based on personal popularity or physical stature—was all that could have mattered to me.

Remembered Lessons

1. Avoid fighting bigger boys or dogs

As a child I lived with being punier than other boys in class. The only consolation was my parents' empathy—they encouraged constant trips to the local drugstore for chocolate milk shakes to fatten me up. The shakes made me happy, but still all through grammar school other kids shoved me around. At first I responded with my fists, but soon I realized that being called a sissy was a better fate than being beaten up. It was easier to cross to the other side of the street than come face-to-face with loitering menaces with a nose for my fear. Likewise, I was no match for barking dogs, particularly ones I had provoked by climbing over fences into their domains. Spotting a rare bird is never worth the bite of a cur. Once bitten by a German shepherd, I knew that I preferred cats, even if they are bird-killers. Life is long enough for more than one chance at a rare bird.

2. Put lots of spin on balls

I long wanted to be part of the softball games played on the big vacant lot across Seventy-ninth Street. At first my only way to join in was to field foul balls. Then I learned how to put spin on underhanded pitches that kept even the better batters from routinely smacking line drives through holes in the outfield. From then on I felt much less an outsider on Saturday mornings. The spins that came from similarly slicing ping-pong serves helped make me a good player well before my arms got long enough to reach near the net of our family's basement table.

3. Never accept dares that put your life at risk

Seeing classmates dash across a street to beat a coming car filled me with more horror than envy of their bravado. When I rode my bike three miles to the Museum of Science and Industry, I knew my constantly worrying mother would have preferred my taking the streetcar. But by being cautious—going down as many alleys as possible and never taking my hands off the handlebars when a car was passing—I was never really putting my life at significant risk. Likewise, in climbing up and over the branches of neighborhood trees or hoisting myself up along gutters to the roofs of one-story garages, I may have been risking a broken leg but not a fatal fall. The possibility of plunging more than ten feet never seemed worth the thrill of being high up.

4. Accept only advice that comes from experience
as opposed to revelation

Listening to my elders just because they were older was not the way I grew up. Preadolescent exposure to my relatives' views that the New Deal would bankrupt the United States and that Hitler would cease being an aggressor after conquering England left me with no illusions that adults are less likely than children to utter nonsense. For the most part, my parents tried to provide rational explanations for why I should think a certain way or do a certain thing. So I was convinced by my mother's advice that I wear rubbers on rainy days so as not to ruin my leather soles. At the same time, I rejected her no less often heard argument that sodden feet led to colds.

By then I was conditioned to accept my father's disdain for any explanations that went beyond the laws of reason and science. Astrology had to be bunk until someone could demonstrate in a verifiable way that the arrangement of the stars and planets affected the course of individual lives. Equally improbable to Dad was the idea of a supreme being, the widespread belief in whose existence was in no way subject to observation or experimentation. It is no coincidence that so many religious beliefs date back to times when no science could possibly have accounted satisfactorily for many of the natural phenomena inspiring scripture and myths.

5. Hypocrisy in search of social acceptance erodes
your self-respect

My parents and most of their neighbors had nothing bonding them together but Horace Mann Grammar School. Mother, with an outgoing and generous personality, naturally rose to be president of the PTA. But except for a keen interest in baseball, Dad had nothing in common with his fellow fathers. That love, however, seldom drew him into the backyards of neighbors, where frequent blasts at the New Deal and occasional anti-Semitic jokes were insufferable for Dad, whose favorite radio personality besides Franklin Roosevelt was the Jewish intellectual Clifton Fadiman. He knew enough to avoid occasions where polite silence in response to repulsive remarks could be construed as acquiescence in their awfulness.

6. Never be flippant with teachers

My parents made it clear that I should never display even the slightest disrespect to individuals who had the power to let me skip a half grade or move into more challenging classes. While it was all right for me to know more about a topic than my sixth-grade teacher had ever learned, questioning her facts could only lead to trouble. Until one has cleared high school there is little to be gained by questioning what your teacher wants you to learn. Better to memorize obligingly their pet facts and get perfect grades. Save flights of rebellion for when authority does not have you by the throat.

7. When intellectually panicking, get help quickly

Occasionally I found myself nervously distraught, unable to repeat an algebraic trick I had learned the previous day. I never hesitated in such circumstances to turn to a classmate for help. Better for one of them to know my inadequacies than not to be able to go on to the next problem. "Do it yourself or you'll never learn" may have some validity, but fail to get it done and you'll go nowhere. Even more frequently I was unable to express myself in words and habitually procrastinated with writing assignments. It was only with my mother's last-minute help that I punctually submitted a well-written eighth-grade paper on the history of Chicago. Of much greater importance was Mother's later insistence that she edit every word of my scholarship essay to the University of Chicago. I accepted her extensive editing with little guilt, then or since.

8. Find a young hero to emulate

On one of our regular Friday night visits to the Seventy-third Street public library, my father encouraged me to borrow Paul de Kruif's celebrated 1926 book, *Microbe Hunters.* In it were fascinating stories of how infectious diseases were being conquered by scientists who went after bad germs with the same tenacity as Sherlock Holmes pursuing the evil Dr. Moriarty. Some months later I brought home *Arrowsmith,* in which Sinclair Lewis, helped by Paul de Kruif as expert consultant, relates the never-realized hope of his hero to save victims from cholera by treating them with bacteria-killing viruses. The protagonist's youth gripped me and made me realize that science could be like baseball: a young man's game whose stars made their mark in their early twenties.

Also encouraging me to aim high was my not-too-distant cousin Orson Welles, whose grandmother was a Watson. Though we never met, he also had an Illinois background and after being effectively orphaned was partly raised by my father's uncle, the celebrated Chicago artist Dudley Crafts Watson. Always turned out with much panache, including a pince-nez, Dudley relished telling his nephew's family of Orson's triumphs, which began when he was a child actor in the Todd School. Orson's daring was what appealed to me most, from his famous *War of the Worlds* radio hoax to his groundbreaking feature *Citizen Kane.* A scientist's hero need not be a microbiologist, let alone a baseball player.

2. MANNERS LEARNED WHILE AN UNDERGRADUATE

I **WENT** to my first college classes at the University of Chicago in the summer of 1943. By starting in the summer and continuing in residence during subsequent summers, I had a good chance of obtaining my degree before I could be called into military service when I turned eighteen. Initially I had no choice about the courses I took—one-year surveys in the physical sciences, the humanities, and the social sciences were the intellectual blue plate special for all freshmen. There were even more prosaic requirements in math and English (reading, writing, and criticism). The survey requisites were a reaction against the free elective curricula that had come to dominate American colleges in the early twentieth century, particularly following the popularization of this system at Harvard by its then president, Charles Eliot.

At the time, the College of the University of Chicago saw as its clear purpose to perpetuate the common ideas and ideals that held together Western civilization. To do so President Robert Hutchins required the college to uphold before students "the habitual vision of greatness." When I matriculated, Hutchins was forty-four. He had become president fourteen years before in 1929 at the age of thirty. He had earlier served as secretary of the Yale Corporation at the age of twenty-four, under James Rowland Angell, who had come from the University of Chicago to be Yale's president. Upon obtaining his law degree, Hutchins began teaching law and through his personal magnetism and confident intellect quickly dominated the Yale law faculty, soon becoming its youngest dean ever. He remained only a year in this prestigious position before being chosen as the sixth president of the University of Chicago.

An impulse to reform the chaotic state of American undergraduate education actually predated Hutchins's arrival in the form of a faculty report recommending that all students take a common set of introductory survey courses during their freshman and sophomore years. Afterward they would take elective courses in their fields of concentration. When he launched this program in 1931, Hutchins grafted onto it two much more radical ideas. The first was the replacement of conventional textbooks with readings from the great books of Western civilization starting with Plato and going through Darwin, Marx, and Freud. Equally revolutionary was Hutchins's plan to accept students after only two years of high school. This idea was implemented experimentally beginning in 1937, largely with students in the University High School and taught mainly in the high school's classrooms. By 1942, however, a close vote of the war-depleted faculty realized Hutchins's bold alternative to the conventional bachelor's degree.

It was into this essentially untried educational environment that I entered each day via a roughly thirty-minute streetcar commute for a three-cent student fare. My best course was Social Science I (American Political Institutions), then taught very ably by Robert Keohane. There was no deep metaphysics on which to get hung up, and I went with pleasure to the main reading room of Harper Library to find primary historical documents such as the Federalist Papers or the *Dred Scott* decision. The book that influenced me most, however, was *Main Currents in American Thought* by Vernon Parrington. It was the first to push me above the canned versions of American history, emphasizing names, dates, maps, and tables to reckon with economic and religious determinism. Much more clearly than before, I appreciated the ideological differences between the Democrats and Republicans and their respective alternatives for coping with the Great Depression, which now only a major war, it seemed, could bring to an end.

Inserting the great books into the science surveys was from the start a controversial idea totally opposed by the science faculty, who considered teaching the history before the facts of science a lunacy of mad medievalists. My introduction to physical science survey, taught by the biologist Tom Hall, was a hodgepodge of these two approaches. Much of the time I couldn't tell what was required of me, and my self-esteem fell when I received a B on my exam at the end of the summer term. Fortunately, only the results of the comprehensive exam taken at the end of the full year's work would appear on my official record. But I got a B on that too.

The evaluation system at the college was then unique. Hutchins had nothing but contempt for the custom of courses being continually punctuated by exams requiring modest recall of textbook readings or lecture notes. At Chicago a special board of examiners, not the individual course instructors, was responsible for the exams. No advantage could come of buttering up the instructor or religious note taking in lectures. Your attention could focus on intellectual arguments while you were in class, not afterward in preparation for an exam. Unfortunately, some of the exams felt more like rarefied IQ tests than honest attempts at evaluating knowledge of the syllabus.

As a commuting student, I entered only marginally into the social life of Hutchins's college, about half of whose students lived in dorm rooms set aside for them. Ida Noyes Hall, originally the social and athletic center for women, became the meeting point for my younger cohort, many of whom, despite relative youth, relaxed by playing endless hands of bridge there. Our athletics also centered on Noyes Hall, where the gymnasium was used for intramural games as well as academic competitions with teams from the private high schools, such as Chicago Latin, the University High School's traditional competitors. I routinely went to all its home court games but more obsessively followed the college team, which in 1943–44 played its last season in the Big Ten. Chicago's compulsory survey courses were ultimately unmanageable for Big Ten–quality athletes, and allowances were not going to be made for students recruited solely for their athletic ability. Our final year was a humiliation until the Chicago Five were vastly strengthened by the arrival on campus of several men from the navy for war-related learning. I became transfixed during our last game of the season, against the perennial powerhouse Ohio State. Chicago kept it tied until almost the end, allowing a tiny crowd of fans to head off to bed knowing they had almost witnessed a miracle.

On the west side of Stagg Field were the original football stands, underneath which handball and squash courts had been placed. I naturally gravitated to handball, in which sheer strength counted for little. Several courts north of where I usually played, there was a locked door with a No Trespassing sign, from which one inferred that war research was being conducted on the other side. I wondered whether it was an extension of the top-secret physics project that had recently brought to Chicago my physicist uncle, William Weldon Watson, who had come with his family from New Haven, where he was a professor at Yale. Though Bill was very discreet, I got the impression that they were trying to develop a superweapon ahead of the Germans.

A real plus of the college's evaluation system was that you could take your comprehensive exams as soon as you felt prepared. No requirements existed for attending classes or writing term papers. And the tuition was the same even if you registered for more than the normal course load. Because of the war, all the second years of physical sciences courses were crammed into the spring 1944 quarter. Once again I eked out a B on the final exam. I used the following summer quarter to cram down the one-year-long Biological Science Survey, which, happily, was not weighted down by the great-books historical approach. With my interest in birds drawing me toward a career in biology, I was disappointed when I got yet another B on the comprehensive exam that August.

My progress toward a concentration in science did not reflect any dislike of the second-year surveys in the humanities or social sciences. In fact, both these classes left lasting memories of inspired teaching. Of all my instructors, the Trinity College–trained Irish classicist David Greene would bring me closest to Hutchins's idea of great teaching. Particularly moving was Greene's Humanities II lecture on the grand inquisitor of Dostoevsky's *Brothers Karamazov* and the choice between freedom and the security offered by adherence to religious authority. I was also captivated by my Social Science II lectures, interspersed with discussion sessions led by the German-born refugee from Nazism Christian Mackauer. With his Continental background he was much at home pitting Max Weber's *The Protestant Ethic and the*

Spirit of Capitalism against R. H. Tawney's *Religion and the Rise of Capitalism.* It was a compelling new outlook on my father's Protestant heritage.

In my first departmental science class, Botany 101, I was several years younger than the other students. The basics of plant anatomy and physiology were little more than an exercise in short-term memory. But the laboratory sessions were a horror since they demanded sketching what I saw under the microscope. My inability to draw, much less do so neatly, depressingly ensured that my final grade would be another B. Zoology 101, taught by the termite specialist Alfred Emerson, went much better: less drawing, and many trips to the Field Museum to look at its extensive collection of reptiles, birds, and mammals.

I remained all through my college years a fervent ornithologist, especially during the spring and fall migrations, when I frequently went by myself, sometimes extending the reach of public transportation by hitchhiking, to prime birding areas. The birds that fascinated me the most were the shorebirds, ranging from the tiny sandpipers to the much larger curlews. I was always on the lookout for the very rare red Wilson and northern phalaropes that Dad had seen when he was a boy. So I was tremendously thrilled when one day in early May, in a marsh on the west shore of Lake Calumet, I spotted three northern phalaropes spinning in the shallow water.

In the spring of 1945, I took the intellectually challenging physiology course taught by the clever Ralph Gerard, whose recent book, *Unresting Cells,* was one of our texts. The class was given in Abbott Hall, which was next to Billings Hospital and was the headquarters of the biochemistry and physiology departments. No longer did lab work depend on drawing. Instead we did actual experiments on frogs whose consciousness had been destroyed by the quick insertion of a sharpened metal rod into their brains. On other afternoons, teaching assistants did demonstrations on anesthetized dogs that had been brought down from the animal room on the top floor of Abbott. In summer months when the windows were open, the sounds of barking dogs reached the walks below, upsetting those who believed experimenting on animals to be morally irresponsible. In contrast I, like almost everyone I knew, saw no alternative to animal experimentation if we were to advance science and medicine.

The spring term was emotionally overshadowed by the death of Franklin Roosevelt on April 12 and the end of the war in Europe less than a month later. Hutchins saw V-E Day, the end of the war in Europe, as an occasion for a major statement and assembled the student body on the morning of May 8 in Rockefeller Chapel. Like many of my friends, my thoughts were of dismantling forever the German military machine that had inflicted on humanity two catastrophic world wars. Hutchins, however, majestically warned against lawless revenge that would go against the ideals for the sake of which we had entered the war. Given the day, the speech was an extraordinarily brave gesture that made me ashamed of my having espoused Henry Morganthau's proposal to reduce Germany to a nonindustrialized, pastoral country.

Academically this was my best term, the first one in which I received two A's—one in physiology, the other in my first advanced divisional course, Botany 234, Physiographic Ecology. This latter course, taught by Charles Olmstead, was a walkover devoted to elucidating differences in plant life as a function of the environment. I spent that summer on a tiny island just off Door County, the long thin peninsula that separates Wisconsin's Green Bay from northern Lake Michigan. With Japanese power in full retreat and the war likely to end soon, there was no longer reason to hurry my education with summer schooling. I opted for a camp counselor position that would bring me into real northern wilderness, away from the oppressively humid heat of most Chicago summers. Though I was underqualified, being neither a strong swimmer nor an experienced boatman, the proprietor was sorely in need of staff, and I became the camp's first "nature" counselor.

Despite being so out of place, I enjoyed most days, slipping away whenever possible into the dense tangle of spruce and fir trees that surrounded the campgrounds. I could walk the perimeter of the island in less than a half hour, ever hopeful that a rare shorebird would fly by. One day this wish was royally granted when three majestic Hudsonian curlews flew within twenty feet of my observation spot. In mid-August the radio brought news of the first atom bombs having been dropped on Japan and the immediate end of the war; it was proof of the superweapon concept, which had brought my uncle Bill to the University of Chicago and its Ryerson Physical Laboratory. Later I eagerly read the *Chicago Tribune*'s detailed account of the University of Chicago's key contribution, the first sustained nuclear reaction produced by man; it had been accomplished in the atomic pile constructed by physicists Enrico Fermi and Leo Szilard, underneath the west football stands where I played handball.

Upon returning to school for the fall 1945 quarter, I decided to risk losing my scholarship aid by taking more difficult courses. Almost all my choices were quantitative, as I simultaneously took calculus, chemistry, and physics. Balancing chemical equations was only mildly painful, and I received two A's and a B. But I was much less at home with calculus, getting a B in differential calculus and then a C in the next quarter's integral calculus. So I took no further math to give more attention to my physics course. Though the instructor, Mario Iona, took a seemingly perverse pleasure in penalizing us for wrong guesses on his multiple-choice physics quizzes, I hit my stride by the spring term and pulled my grade up to an A.

All through the year I was taking OII (Observation, Interpretation, and Integration), the most philosophically oriented of all the surveys and the last required course I needed for my degree. Again I had good fortune regarding the instructor: Joseph Schwab, whose languorous southern tone could never hide his disdain for crap answers to precise Socratic questioning in my Humanities I class. Now that I had learned to anticipate his line of interrogation, I quite often enjoyed and even looked forward to his Swift Hall classes, particularly after getting past medieval thought and on to that of the Renaissance and Francis Bacon's distinction between deductive and inductive reasoning. Even more appealing were the masterfully clear writings of the late-nineteenth-century Harvard logician C. S. Peirce. In the end, however, I was brought back to reality by my B on the final comprehensive. The message was clear that I should avoid further philosophical inquiry.

By dropping the last quarter of math, I had time to audit lectures on physiological genetics by Sewall Wright, the university's best-known biologist. Besides being one of the world's most accomplished population geneticists and a leader in evolutionary thought, Wright also did much to advance biology toward understanding what genes do at the biochemical level. I had by then independently become focused on the gene through a reading that winter of Erwin Schrödinger's thin book *What Is Life?* Because Schrödinger, the inventor of quantum wave mechanics and a 1933 Nobel Prize–winning physicist, had seen the importance of writing about biology, *What Is Life?* was featured in the Sunday book section of the *Chicago Sun-Times*. The next morning I checked it out of the biology library. Schrödinger elegantly laid out how genes were the most important feature of life, since they maintained its continuity by carrying hereditary information from one generation to the next. As birds had bound me to life sciences, Schrödinger's exaltation of the gene would lead me to a life of studying genetics.

At the June 1946 commencement, Robert Hutchins elegantly warned of the doom of Western civilization unless university graduates led the world toward a moral, intellectual, and spiritual revolution that contrasted with today's postwar "get all you can" ethos. I was struck by our president's tone that morning—not that of the metaphysician but that of the preacher's son—as he fervidly admonished us that only by coming to see our fellows as the children of God would we stop seeing them as rivals. He further warned us that unless we esteem ourselves as more than animals we are doomed to act like them, and the laws of the jungle will prevail. Most unexpected were his statements that we can practice Aristotelian ethics only with the support and inspiration of religious faith, that the brotherhood of men must rest on the fatherhood of God, and that cats and dogs are more attractive than most humans.

Excited by his forceful rhetoric but uneasy about its sentiments, my parents and my sister, who had just completed her first year at the college, walked across Woodlawn Avenue to the reception at Ida Noyes Hall. There Hutchins recognized Dad as we passed through the receiving line, and briefly chatted with him about their student days at

*My advanced ornithology class at the University of Michigan Biological
Station, summer of 1946. I am in the back row, second from the left.*

Oberlin. Afterward Dad recounted how Bob had then been a rebel and
had been part of a group that secretly smoked cigarettes.

At the start of summer I was on the train to Pellston, just below the
Straits of Mackinac. Nearby on Douglas Lake was the University of
Michigan Biological Station. Upon my arrival I registered for two
courses, Systematic Botany and Advanced Ornithology, moved into a
tented cabin, and soon found myself socializing mainly with those,
like me, who were serving meals in the dining hall to make our limited
means go farther. Now more than six feet tall, I no longer looked a
physical misfit. For the first time I began to befriend peers who were
not obvious oddballs, elected because no one else seemed keen to eat
with me. Soon I was being called "Jimbo," a southernism quickly
adopted by several young waitresses who were inspired by my youth to
treat me as a kid brother.

Upon my return to fall classes, my seemingly improved social skills
encouraged a close friend and his girlfriend to set me up on my first

"Jimbo" Watson,
eighteen years
old and at ease

real college date. Dominating the college social scene during my junior and senior years were two girls who were almost always seen together: Rosemary Raymond and Irene (Reno) Lyons. Rosy was the more smashing, often spotted driving about in her parents' new, Raymond Loewy–designed, ultramodern Studebaker. But Reno was also clearly fun, and I did not believe she would go out with me. She agreed, however, and we went to a Saturday night party in a campus dorm, from which we went on to have a snack at the Tropical Hut on Fifty-seventh Street, where my awkwardness abated only somewhat.

That was not only my first college date but also my last. The pursuit of happiness thereafter was mainly taken up with improving my tennis skills on the courts under Stagg Field with Howard Holtzer, my only real competition in zoology. Though he was five years older than me, I could occasionally beat him with cross-court forehand shots. At that time I was just filling out several applications for graduate school, with Caltech as my first choice. Not only was its Biology Department heavily into genetics, but its world-famous chemist, Linus Pauling, was also interested in biology at the molecular level. I also applied to the Biology Department at Harvard for no good reason except that Harvard was Harvard. Indiana University was a sleeper that I applied to on the advice of my undergraduate advisor. I was told they had several outstanding younger geneticists, though their names then meant nothing to me. I did know, however, that the great geneticist Hermann J. Muller had just arrived there. He became a celebrity that October when it was announced he had won the 1946 Nobel Prize for his discovery in 1926 that X-rays cause mutations.

As early April approached, I began to worry about the fate of my applications. And so it was a considerable relief when I did receive an acceptance letter from Fernandus Payne, the dean of Indiana University's graduate school. In addition to a $900 annual stipend, all tuition fees were waived. I was warned, however, that if I expected my main interest to continue to be birds, I should choose a different place. So my future was already secure when I received Caltech's rejection letter, which hurt but didn't surprise me. They had no way of knowing that I had become more interested in studying genetics than birds. They also expected their students to do well in math. Last to arrive was a letter from Harvard offering acceptance but no financial aid for tuition or living costs. I was in no sense disappointed, as there was no one on its faculty truly interested in the gene.

During my last quarter, I registered for the simpler of the two departmental organic chemistry courses, the one for pre-meds as opposed to future scientists. The course material was not much of a challenge, and the professor gave us the option of counting only the top three out of four required exams toward our final grade. After getting A's on the first three tests, it seemed perfectly reasonable to skip the final two and a half weeks of lectures. In this way, I graduated from Chicago never having studied ring-shaped, aromatic carbon compounds. My grades for my two other spring courses, which included

INDIANA UNIVERSITY
BLOOMINGTON, INDIANA

GRADUATE SCHOOL
OFFICE OF THE DEAN

March 31, 1947

Mr. James Dewey Watson
7922 Luella Avenue
Chicago, Illinois

My dear Mr. Watson:

It is a pleasure to notify you that you have been recommended for an All-University Fellowship in Zoology for the academic year 1947-48, at a stipend of $900 for the period mentioned. While your appointment has not been officially approved, the administration has given us permission to give you this unofficial notice.

Will you please let us know at your early convenience whether or not you will be able to accept the appointment?

Very truly yours,

Fernandus Payne
Dean

FP:la

P.S. If your interests continue to be in ornithology it would be my advice to refuse this offer for we make no pretense of offering graduate work in that field.

F Payne

Indiana University made its position perfectly clear.

Vertebrate Embryology and a statistics-oriented psychology course, were also A's, as in fact were all my grades the final year. I ascribed my success in the final laps primarily to lack of competition, since better science students tended toward the more rigorous programs of physics and chemistry.

Just before the June 1947 commencement, at which I would receive a B.S. degree, I learned that I had been elected to Phi Beta Kappa. I had long coveted the honor but never thought I could pull off enough A's in science to pull up the rest of my only-slightly-above-average grades. Most thrilled were my parents, who had devoted so much of their meager salaries to my schooling. Never once did they even hint that I was shirking my responsibilities in preferring books and birds over making money.

Over my last summer at home, I spent more and more time at Wolf Lake observing shorebirds. There I found hundreds of black-bellied plovers as well as dozens of the golden plovers, whose long migration over oceans had so thrilled me when I read about them as a child. Piping plovers still nested on the sandy shores and intermingled with the many more semipalmate plovers that arrived soon after the fall migration started in early August. I would watch the shorebird flocks scurrying along the beach, trying to think through the evolutionary pressures that had created such social animals, which invariably flocked together instead of going their solitary ways. Whether any general principles governing the social behavior of birds even existed, no one then knew. Deep down I was relieved that Indiana's warning would not allow me to stake my future on the pursuit of an objective that might prove a phantom. In contrast, what needed to be found out about genes was relatively clear. They obviously existed, but how did they work? Knowing by graduation where I wanted to go intellectually was the real achievement of my college years.

Remembered Lessons

1. College is for learning how to think

Whether on a scholarship or paying full fare, college costs too much time and money if you don't use it to learn how to think. In Joseph Schwab's Humanities I class, knowing what Socrates was reputed to have said mattered much less than confronting whether the reasoning he used to reach his conclusions was watertight. Day after day we were pushed into classroom fights where we boxed with our brains instead of our fists. Syllogisms, in which two premises predicate unassailable conclusions, dominated our classroom hours, with the exact meaning of words being of paramount importance. Failure to spot faulty premises all too easily led to classroom answers that ran against common sense. Back in high school, I was challenged by a friend who had what he thought was incontrovertible proof of the existence of God.

Going back home, I felt stupidly unable to fight back against his word war, even while suspecting that he was pulling a fast one that I was too dumb to spot. Later when we both went to the University of Chicago, he came up against much better practitioners of semantics and logic. Soon Bill sheepishly confessed to me that syllogisms no longer led him to God, and freed from thinking about sin, he developed a new preoccupation with girls—and more girls.

2. Knowing "why" (an idea) is more important than learning "what" (a fact)

World Almanac facts, such as the relative heights of mountains or the names of British kings, got you nowhere at Hutchins's college. The essence of its educational mission was the propagation and dissection of ideas, not the teaching of facts often best left to trade schools. Why the Roman Empire had risen and fallen was much more important than the birth date of Julius Caesar. And why the great European cathedrals were built mattered much more than their relative sizes. Equally unimportant were the details about the French Revolution when contrasted to the philosophical ideas of its eighteenth-century Enlightenment, whose emphasis on reason as opposed to theological revelation greatly accelerated the development of modern science. Likewise, details of Linnean taxonomy paled in significance to the idea of biological evolution, whereby all life-forms have a common ancestor. Better simply to know which books hold details you will need than to overload your neurons with facts that later will never need to be retrieved.

3. New ideas usually need new facts

Though facts are inherently less satisfying than the conclusions drawn from them, their importance cannot be denied. Darwin's abandonment of the Bible's description of the origin of life came out of the observations that he made as a naturalist aboard the HMS *Beagle.* During its three-year mission of charting poorly known stretches of the western coastline of South America, Darwin spent much time ashore observing and collecting the fauna, flora, and fossils. He noticed among other things that species found in mountainous temperate regions were more similar to those found in nearby tropical regions than they were to species from temperate regions on other continents. Likewise, many South American fossils appeared more related to living South American equivalents than to fossils found, say, in Europe. Crucial to his own acceptance of the origin of species by natural selection was his visit to the Galapagos Archipelago, where each island had its own unique species of finches. Effectively isolated from interbreeding with finches of neighboring islands, each species evolved its own unique coloration and beak shape. Sometimes a new idea can flow from old facts rearranged, but more typically it comes when new things previously unknown and unaccountable for under the old theory are introduced.

4. Think like your teachers

Learning to think should also make your life easier. During my first university years, I crammed far too much for exams, trying to be on top of all the topics given even semiprominence in my syllabi or texts. It would have been much better to focus on questions my teachers were certain to ask, which I could discern if I paid attention to their main take-home lessons. Trying to put myself into my instructor's head became much easier when I began to concentrate on subjects of personal interest, and I proved a whiz while a senior at mastering the ideas of ecological animal geography.

5. Pursue courses where you get top grades

After you've satisfied requirements, choose courses that naturally interest you, not ones someone else thinks you should care about. Then give these courses your all. If your grades in classes you like are not largely A's, you have likely not yet found your intellectual calling. As a corollary: if you do take courses that prove to be no fun, don't get upset by less than stellar grades.

6. Seek out bright as opposed to popular friends

Though big-time sports were gone, the University of Chicago still had fraternities that acted as dining and housing centers for students with social aspirations after the war ended. Only one such meal in a house on University Avenue made it obvious to all that I'd better plan to continue dining at the Hutchinson Commons with several undergraduate science oddballs who, like me, could not generate polite words of no purpose. There we could frequently look across its long refectory tables and see Enrico Fermi talking with his graduate students and postdocs. The great Italian-born physicist had elected to be there instead of dining with fellow physics professors in the more stuffy Quadrangle Club. In my senior year, I began moving among zoology department graduate students. With them small talk came easily. I didn't feel like an oddball, as I did among students of other aspirations who were soon to move off into worlds of which I had no interest in ever becoming a part.

7. Have teachers who like you intellectually

Long after I left Chicago, a gathering of its New York alumni brought me into contact with a classmate whom I remembered best for his devotion to bridge games. Cheerfully he told me how, given my social awkwardness, none of my classmates thought I would amount to much. Happily it was my teachers' opinion, not my classmates', that mattered. After fun classes I liked to stay on to ask questions, and in this way I became well known to them. Because of my enthusiasm, during my senior year the animal behaviorist Clyde Allee invited me to join the weekly meeting of his students at his nearby Hyde Park house. He not only gave me an A but also wrote a compelling recommendation for my application to graduate school. The most effective endorsements are cultivated well before application time. So it was with the human geneticist Herluf Strandskov, who would also praise my keen interest in biology. You don't have to win them all: the impervious embryologist Paul Weiss was ever annoyed at my failure to take notes during his lectures while still getting A's in his exams. I had the sense not to ask him for a recommendation.

8. Narrow down your intellectual (career) objectives while still in college

The University of Chicago was virtually an officers' training school for intellectuals, and its Zoology Department had no rival in range elsewhere in the United States. Though initially excited by virtually all aspects of biology, by late in my junior year I found myself keenly interested in the gene. If I had not figured out my focus so early, I very likely would have gone to a school such as Cornell or Berkeley that had great programs in biology but not in genetics. In such circumstances I might have grown bored with my thesis research and been obliged to wait until after my Ph.D. was completed, some three or four years, before experiencing true intellectual excitement. And by then I would have left the most thrilling problem of all—the DNA structure—for others to solve.

3. MANNERS PICKED UP IN GRADUATE SCHOOL

INDIANA University (IU), at Bloomington in Monroe County, to which I went in September 1947 to become a scientist was waking up from a genteel past that made it still more redolent of a southern state university than the other earnestly progressive Big Ten partners such as Wisconsin and Michigan. Presiding over the emergence of IU as an institution where learning and research were to be as important as weekends of fraternity- and sorority-led fun was Herman B. Wells. He became president in 1937, when he was only thirty-five. Short and heavyset, the Indiana-born small-town boy had prodigious energy and visions of greatness for his university, until then the academic runt of the Big Ten.

During its more than thirty-year slumber since 1910, President William L. Bryan had not presided over the construction of even one major new dormitory. So Wells moved quickly to shake things up, recruiting new faculty and using Depression-fighting federal funds to build important new facilities, including a grand four-thousand-seat student auditorium. For science faculty recruiting, Wells tapped Fernandus Payne, who until then had been badly underutilized by the graduate school. A Hoosier born and bred, Payne, like almost all the senior faculty, had gone east to earn his Ph.D., which was awarded in 1909 by Columbia University, where *Drosophila*, the tiny fruit fly, had just been introduced into the laboratory of T. H. Morgan. Payne had the bad luck to return to Indiana just before Morgan and his students Alfred Sturtevant and Calvin Bridges began their seminal experiments, mapping genes to fixed locations along the *Drosophila* chromosomes.

Though never a major player in genetics, Payne knew where the action was. He made offers to major talents not yet adequately discovered or recognized by major institutions. From Goucher, the women's college in Baltimore, he recruited the respected cytologist Ralph Cleland in 1938 for the Botany Department. A year later he brought to the Zoology Department the extraordinary protozoologist Tracy Sonneborn from Johns Hopkins. Then late in 1943, the bacteriology department acquired the Italian-born Salvador Luria, initially trained as an M.D. but by then exploring the genetics of viruses. Even more spectacular was Payne's ability in 1946 to persuade the already world-famous *Drosophila* geneticist Hermann J. Muller to join IU's ranks.

In appointing Sonneborn and Luria, Payne took no notice of the Jewish heritage that had kept Sonneborn from a tenured appointment at Johns Hopkins and Luria from an invitation to join the faculty at the College of Physicians and Surgeons of Columbia University, where he had been given a temporary position upon his arrival as a wartime refugee. Muller, then working with Sewall Wright, the most accomplished of American geneticists, suffered from a double whammy. Not only was he Jewish, but his departure from Texas to Moscow in the early 1930s had indelibly marked him as an unemployable leftist. Upon his return to the States in 1940, only by the intervention of his friend H. H. Plough could he secure a temporary position at Amherst. He had asked many other friends for help, but no major research university made him an offer, even though Muller had by then turned totally against the Soviet government. So it was with great relief that Muller accepted the IU professorship. Soon Indiana would have even greater reason to be pleased when Muller was given the Nobel Prize in Physiology or Medicine.

To get to Bloomington I took the Monon, the railroad most identified with Indiana's past and one whose original tracks went past the Gleason farm near La Porte where my grandmother was raised. Along these tracks Nana saw Lincoln's funeral train as it slowly crisscrossed the Midwest toward Springfield, Illinois, where the president was buried. I had signed up to live and eat at the Rogers Center, a post-

Herman J. Muller, 1941

war utilitarian dormitory complex located a mile east of the campus center. Its two-story buildings were of necessity built too quickly to have the permanent elegance that comes with Indiana limestone. My $900 fellowship would more than cover room and board fees, leaving me enough for the occasional movie and meal out.

Some twenty thousand students were then enrolled at IU. All Indiana high school graduates were entitled to enroll there or at rival Purdue, the engineering and agriculture-oriented university a hundred miles north. The state saw its obligation to offer everyone who wanted one a good education. But each year half the freshman class did not go on to be sophomores. Poor grades were behind most dropouts, but a significant number of girls transferred to other schools because of their failure to be accepted by a suitable sorority.

The aged zoology and botany laboratories that dated from the 1880s were in no way adequate for IU's new genetics thrust. Ralph Cleland, upon his 1938 move to the Botany Department, however, had to make do. Tracy Sonneborn had the better fortune a year later of being given space in the relatively new chemistry building. Salvador Luria, coming in 1943, found his new Microbiology Department lab sited in the attic of the early-twentieth-century Kirkwood Hall, originally designed for physics and chemistry, though by then foreign languages and nutrition were taught on lower floors. Hermann J. Muller's lab was hastily created in 1946 in the basement of the equally out-of-date psychology building. As a first-year graduate student, I was given a desk on the top floor of the zoology building, whose original elevator was still operated by pulling ropes to go up and down.

On the first floor of the zoology building were the offices of Alfred Kinsey, much esteemed for his studies on gall wasps, and until recently the teacher of the undergraduate course on evolution. By 1947, however, his focus had turned exclusively to human sexuality, then a daring topic for any university, particularly one almost in the South. Fortunately, Kinsey's recently published book summarizing his findings was so heavily statistical as to be more likely purgative than prurient. In fact, subsequent criticism that Kinsey seemed ignorant of the emotional aspects of sex later led his Institute of Human Sexuality to accumulate a highly restricted library of erotica. This backfired when some books later purchased in France were seized by U.S. Customs and, despite much pleading by Herman Wells, never released for their scholarly aims.

The day after my arrival, I arranged my courses for the coming term. Naturally I signed up for Muller's Advanced Genetics—Mutations and the Gene. I was also urged by Fernandus Payne to take as soon as possible Microbial Genetics with Tracy Sonneborn, since he was zoology's brightest young star. But that term he was only teaching an elementary genetics class, and so I registered for Salvador Luria's course on viruses. Soon I heard faculty gossip that Luria treated his students like dogs. This worried me until I listened to his first several lectures and found them mesmerizing. Less comprehensible to my Zoology Department advisers was my desire to register for Advanced Calculus, a course usually taken only by physics and math majors. But unless I took it, I feared, I would never have the courage to learn more physics, without which I might be precluded from pursuing possible high-powered ways to probe the gene. Ironically, my teacher was to be Lawrence Graves, on sabbatical from the University of Chicago, where I never would have dared enter into one of his courses. But at the more low-key Indiana I would not be competing with real math whizzes— and besides, grades were rather beside the point.

The required text of Muller's course was the lucid and still highly relevant *Introduction to Modern Genetics* (1939) by the English biologist C. H. Waddington. The heart of the course, however, was Muller's lecture account of his career starting from his days as a student in the "fly room" of Columbia University between 1910 and 1915. Emanating from a short, heavyset man almost the shape of a *Drosophila* himself, Muller's lectures were streams of consciousness rather than prepared orations. His agitated speech mingled clever genetic reasoning with details of his frustrations over, say, not initially being accepted into T. H. Morgan's lab, and later when finally a member having his ideas given short shrift. Much less absorbing were the lab sessions, in which we were chaotically run through an increasingly complex set of genetic crosses. The insights of such experiments seemed rather arcane, pointing to a truth that could not be avoided: *Drosophila*'s days as a model organism were over. Indeed, a new one would soon supplant it as the premier tool for studying the gene.

Through Luria's virus course lectures, I saw the genetic wave of the future unfolding. The key would be microorganisms, whose short life cycles would permit genetic crosses to be done and analyzed in a matter of days instead of weeks or months. Luria was particularly excited about the future of research using the common intestinal bacterium *Escherichia coli* and its parasitic viruses, the bacteriophages (or phages for short, as they were more often called). Soon after his 1943 arrival in Bloomington Luria, then thirty-one, was the first to systematically show that both *E. coli* and its phages gave rise to easily identifiable spontaneous mutants. Only three years later, in 1946, was genetic recombination between different *E. coli* strains demonstrated by the precocious twenty-one-year-old Joshua Lederberg, then a medical student in Edward Tatum's laboratory at Yale. The same year Alfred Hershey at Washington University found genetic recombination for the *E. coli* phages T2 and T4 and soon constructed the first genetic maps of phage chromosomes.

Until Luria's first lecture, I had no idea what a virus was. Soon I knew they were very small, infective agents that multiplied only within living cells. Outside of cells, viruses are essentially inert. But once they enter a cell, a multiplication process is initiated that leads to a generation of hundreds to thousands of new progeny viral particles identical to the original parent particle. Unlike bacteria, viruses cannot be observed using conventional microscopes. Their sizes and shapes first became known following the invention of the much more powerful electron microscope in Germany just prior to the start of World War II. The first phages so examined had unexpected tadpole-like shapes, with polygonal heads attached to much thinner, tail-like appendages.

More than two decades earlier, H. J. Muller had speculated that viruses were, in fact, naked chromosomes that had acquired special structures for being transported from one cell to another. Supporting his conjecture was the finding in the mid-1930s that DNA, a soon-to-be-discovered major component of all chromosomes, also was a major component of the phages. Even more important was the 1944 discovery by Oswald Avery and his coworkers at the Rockefeller Institute in New York that DNA could transmit genetic markers in pneumonia bacteria. Conceivably much, if not all, of the genetic specificity of phages also resided in their DNA components.

Luria's lectures were also particularly exciting for me because they frequently described his collaborations of the past six years with the German-born physicist Max Delbrück, whose ideas about the gene in the mid-1930s provided the essence of Erwin Schrödinger's *What Is Life?* How a gene is copied to yield an identical replica was now being extended by Luria and Delbrück to ask how a single phage particle gives rise to hundreds of identical progeny. In learning how phages multiply, Luria and Delbrück thought the fundamental mechanism of how genes are copied would also become known.

A key requirement of Luria's course was the term paper, which I chose to write on the effects of ionizing radiation on viruses. Luria had used X-rays to estimate the size of the then still submicroscopic phages when he worked in Paris in 1938–40, where he had fled when Mussolini, in an attempt to curry favor from Hitler, had begun his first serious persecution of Italy's Jews. Because only a single ionizing event is necessary to kill a phage, a minimal size of a phage can be calculated from the number of phage particles killed as a function of X-ray dose. The so-called target theory approach was previously used in 1935 by Max Delbrück to give an estimate of the size of *Drosophila* genes, and so I had no difficulty finding enough material to fill out my paper, as no original thought was expected. I was more worried I would be penalized for bad handwriting, but I got an A.

Not so easy was my math course, whose text was *Advanced Calculus* by Harvard's David Widder. Fortunately, Graves began to appreciate the much lower math aptitudes of his Indiana students in comparison to those he had been used to teaching at the University of Chicago.

What had threatened to be entirely above my head got easier, even occasionally satisfying toward the end. Helping matters was the presence in the class of a small, neat blonde with whom I could compare homework answers at the IU Union cafeteria. A grade of B was more than encouragement enough to continue the course through the spring term. Being able to pass a real math course was a big step forward for me, not only for its own sake but also for allowing me to hold my own with the growing number of physicists moving toward biology to find the secrets of the gene.

Not at all surprising, but nevertheless satisfying, was an A+ in animal ecology. Teaching it was Lamont Cole, a mathematical ecologist, newly recruited by Fernandus Payne to broaden the fish-dominated ecology outlook of IU. I loved learning details of animal adaptation to their environments as well as taking weekly field trips to observe how remarkably specific were the adaptations of certain species to certain niches. On a trip to one of the limestone caves that peppered the rolling hills near Bloomington, armed with rubber boots and several coal miner's lamps, we squeezed through narrow openings into sometimes vast, waterlogged cavities in search of blind cave fish. In the absence of light, there was no selective pressure against the emergence of mutant fish lacking not only scale pigments but also functional eyes. It was through studying blind cave fish that the Indiana zoologist David Starr Jordan rose to prominence. A scientist of great charisma, he would lead IU before being chosen in 1891 as the first president of Stanford University. By my time at IU, however, Jordan was locally best known for quipping that every time he learned the name of a student he forgot the name of a fish.

IU's traditional preoccupation with fish reflected the presence of thousands of lakes dotting the Indiana landscape, ranging from tiny farmer's ponds to lakes many miles wide whose shores were lined with vacation cabins. When I was a child, my mother's Gleason relatives occasionally took us fishing on lakes near Michigan City where on good days we would hook and later fry more bluegill, perch, and bass than we could easily eat. On other days, we would get no bites and go home bitter. The State Department of Conservation began helping the university support its biological station at Winona Lake, where fish yield was measured in units called "fish pole hour." While there were lakes generally yielding at least several fish per pole hour, there was also the very sad Oliver Lake, where more than ten hours would be needed to bring home a single fish.

By then I was routinely traveling on foot, three times daily, the two miles from my Rogers Center dorm to the science complex and back again. Because of the overcrowding, everyone in Rogers had a roommate and those of us with lab connections tended to avoid our dorm rooms except for sleeping. On these long walks, I liked to go by Jordan Avenue, site of the most desirable sororities, where I would spot girls much prettier than most to be seen in science buildings. For a break from homework or studying for exams, I would occasionally go birding with Palmer Skaar, a fellow new graduate student, who could identify the local birds as well as I. Best of all were the basketball games that began with IU predictably skunking neighboring DePauw. In contrast, most Big Ten games were cliffhangers until the tense closing moments of the last quarter. Also fun, though more intellectually demanding, were the informal Friday night seminars on protozoan genetics that Tracy Sonneborn had begun having at his home to interest the new group of graduate students in his lab's research.

The more I learned about phages, the more I became ensnared by the mystery of how they multiplied, and even before the fall term was half over I knew I did not want to do my degree with Muller. Nor did Muller's work, which seemed increasingly outdated, attract any of the new students his famous presence had drawn to IU that year. Most were captured by Tracy Sonneborn's infectious enthusiasm for the tiny, one-celled, ciliated protozoan *Paramecium*. I, however, could see no way for paramecia to compete with phages in pursuit of the fundamental nature of the gene. I therefore had to tell Sonneborn, somewhat sheepishly, that I would be working with Luria, fearing that this would spell the end of my welcome at his Friday night protozoan soirees. But he very graciously gave no evidence of feeling spurned and told me that I could keep coming as long as I wanted.

As soon as spring term started, I began learning how to infect *E. coli* cultures with the T2 phage and to count the number of bacteria that had phages multiplying within them. To my great benefit, I also became friends with thirty-three-year-old Renato Dulbecco, who, like Luria, had trained as an M.D. in Turin and who had come the previous fall to learn how to work with phages. With his family still in Italy, Renato was almost always in the lab and could give me needed advice when the phage counts were not what I expected.

With a light spring course load, I started assisting in the bird course, where help was clearly needed given the prevailing fish bias. There was no real bird expert among the zoology faculty, and so the course was taught by Bill Ricker, the department's best fish man. Long regarded as a gut course—anyone who went on the field trips could expect a passing C—it was now used by physical education majors in the new academic division called Health, Physical Education, and Recreation (HPER) to fulfill their science requirement. In fact, this division came into existence the year before to prevent a repetition of the worst tragedy in IU's history. Robert Herchenmeyer, the best football player ever to attend IU and who in the fall of 1945 led it to its first and only Big Ten championship, flunked out the following spring. Yet for the sake of appearances, HPER majors had to take several courses with nonathletes. And so I found myself helping two jocks learn to identify birds seen on previous Saturday morning field trips and thereby pass the final. For my trouble, I got a look at a number of birds virtually unseen in Chicago, such as the Kentucky warbler, the Bachman sparrow, and the pileated woodpecker, the most impressive of all the birds of southern Indiana.

Back in the lab my first research problem from Luria was to see whether phages inactivated by X-rays could still undergo genetic recombination and produce viable recombinant progeny lacking the damaged genetic determinants present in the parental phages. Two years earlier, Luria had discovered that phages inactivated by ultraviolet light did so recombine when two or more infected the same *E. coli* host cell. It was to further analyze "multiplicity reactivation" that Luria had brought Dulbecco over from Italy, and he now hoped I could extend his discovery by using X-rays to produce genetic damage. From my first experiment, I began to get positive results, and

Luria would regularly look over my notebooks to convince himself that I had done the experiments correctly.

I had to briefly interrupt my experiments when my turn came to give a formal lecture before the zoology faculty. All beginning students were expected to give one during their first year as practice for later careers likely to involve some form of teaching. Since I had come to IU for a zoology degree, I prepared a talk on birds that summarized the conclusion of *Darwin's Finches,* the new book by the English ornithologist David Lack. Just before the war, he had lived on the Galapagos Islands to follow up Charles Darwin's bird observations of a hundred years before, which had led Darwin to question the immutability of species.

A week after my talk, I first met Max Delbrück. On his way back from a visit at Princeton with his great hero, the Danish theoretical physicist Niels Bohr, he came to spend several days with Luria to learn how the experiments on multiplicity reactivation were progressing. Max was not at all the middle-aged, balding, somewhat overweight German academic I was expecting. Instead my first visit to Luria's flat, several blocks to the south of the main campus, brought me face-to-face with a man who looked more like a fellow student. Max, then forty-one, had come to the United States in 1936, when he was thirty. As a member of the German Protestant academic elite, he was first excited by astronomy. But by twenty, his interests had shifted to theoretical physics, as quantum mechanics was coming into existence. After obtaining at Göttingen his Ph.D. at twenty-four, he spent several years at Copenhagen, the world center for theoretical physics, under Bohr's tutelage. Returning in 1932 to Berlin to work in the great chemist Otto Hahn's Kaiser Wilhelm Institute, Delbrück became acquainted with the Russian-born *Drosophila* geneticist N. Timofeeff-Ressovsky, who was then using X-rays to induce mutations in *Drosophila* with the help of the physicist K. G. Zimmer. Discussions among the three of them led to their seminal 1935 paper "On the Nature of Gene Mutation and Gene Structure," whose ideas formed the core of Erwin Schrödinger's *What Is Life?*

To move up the German academic ladder, Delbrück needed to demonstrate ideological correctness to the Nazi bureaucracy. Failing to do this, in the fall of 1936 he seized upon an offer from the Rockefeller Foundation of a fellowship that would let him work at Caltech. There T. H. Morgan and his now also famous coworkers, Alfred Sturtevant and Calvin Bridges, had moved in 1929. But after his arrival in Pasadena, Delbrück found *Drosophila* boring and instead turned to working with the physical chemist Emory Ellis on phages in the basement of the biology building. When the war came, as a German alien he was not eligible for war research. Instead, with continuing Rockefeller support, he moved to Vanderbilt University as an instructor of physics. His collaboration with Luria began soon after Luria's arrival in New York City from Italy. During the summers of 1941 and 1942, they did research at the Biological Laboratory at Cold Spring Harbor on Long Island's North Shore. During the following winter months, Luria came temporarily to Nashville as a Guggenheim Fellow before assuming his position at IU early in 1943.

At supper, Delbrück and Luria talked fondly of their collaboration, particularly at Cold Spring Harbor. Earlier Luria had told me that he would be spending the following summer there and invited Dulbecco and me to accompany him. Delbrück also would be returning, as he had done every summer since 1941. I listened attentively to talk about the summer course on phages that Delbrück had started in 1945. The summer before, the course had attracted Leo Szilard, the Hungarian physicist, whose name would be forever linked with Enrico Fermi's for their creation of the first sustained nuclear reaction at the University of Chicago in 1942. Luria and Delbrück spent the day after that dinner talking about phages before playing doubles tennis with Dulbecco and me. I became acutely aware that I would have to elevate my game greatly if I was ever to play singles with Delbrück.

Summer had effectively arrived in Bloomington, with the late May daytime temperatures in Luria's inadequately air-conditioned attic lab often too high for our agar plates to gel quickly. Doing more experiments that term made little sense, even when the temperatures temporarily dipped, and besides, I was preparing for my math finals. I wasn't worried about Advanced Calculus. Graves had let us know that all graduate students would get at least a B for staying in the course (by

Max Delbrück with Salvador Luria and his coworker
Frank Exner at Cold Spring Harbor symposium, 1941

then all the undergraduates had dropped out). But Differential Equations with Professor Kenneth Williams was another matter. Everyone knew that his real passion had long since become the Civil War and his books on the topic, for which he had received much acclaim. I had no rapport with him, finding him mean-spirited when not boring, and was relieved to get off with a B–.

I was in high spirits when I briefly returned home on my way to Cold Spring Harbor and my first trip to the East Coast. The year had exceeded even my highest hopes, since I was now in the thick of the quest to understand the gene. I became more than aware of the advantages of having attended the University of Chicago, where I had learned the need to be forthright and call crap crap. It was not that I was inherently brighter than my fellow graduate students; I was just much more comfortable challenging ideas and conventional wisdom, whether it concerned politics or science. And never forgetting Robert Hutchins's wisdom that the academic world abounds in triviality, I was ever mindful of what sort of work would further my career and what would be merely idle learning.

Remembered Lessons

1. Choose a young thesis adviser

The older the scientist you choose to do your Ph.D. thesis with, the more likely you will find yourself working in a field that saw its better days a long time ago, possibly before you were born. Even when a mature scientist still has all his marbles, he often wants to put more bricks into an edifice that already has enough rooms. This was most certainly the situation with H. J. Muller when he came to Indiana. Though his lectures had much to offer graduate students, the heyday of research like his had long passed, which meant there were no obvious good job prospects for those now working with him. Young professors in contrast are generally hired not for grandeur but because they represent a new intellectual thrust not present in a department, one with hopes of remaining lively over at least the next decade. Moreover, they are likely to have smaller research groups than more senior professors, around whom funds as well as stodgier minds tend to aggregate. I most certainly profited from being Salva Luria's first Ph.D. student and not having to share his attention with other students. Then just expecting his first child, Luria did not yet have the family man's demands on his time, and so even on weekends he was frequently in his lab office, his work proceeding as fast as humanly possible.

2. Expect young hotshots to have arrogant reputations

Those not in Luria's immediate circle frequently made snide remarks about the air of superiority evinced by the phage group around him and Delbrück. Luria's occasional brusqueness in dismissing a scientific objective as rubbish not worthy of a seminar was bound to unsettle colleagues accustomed to good Hoosier manners, whereby good neighbors didn't judge one another too critically. But no matter how polite, intellectuals who break new ground inevitably threaten minds continuing in old ways. Believing that your way to success holds more potential than past approaches and not saying so is of no service to your students. But outside their devoted coteries, intellectual pioneers are bound to be found arrogant at best, delusional at worst. Use your head and draw your own conclusions.

3. Extend yourself intellectually through courses that initially frighten you

All through my undergraduate days, I worried that my limited mathematical talents might keep me from being more than a naturalist. In deciding to go for the gene, whose essence was surely in its molecular properties, there seemed no choice but to tackle my weakness head-on. Not only was math at the heart of virtually all physics, but the forces at work in three-dimensional molecular structures could not be described except with math. Only by taking higher math courses would I develop sufficient comfort to work at the leading edge of my field, even if I never got near the leading edge of math. And so my B's in two genuinely tough math courses were worth far more in confidence capital than any A I would likely have received in a biology course, no matter how demanding. Though I would never use the full extent of analytical methods I had learned, the Poisson distribution analyses needed to do most phage experiments soon became satisfying instead of a source of crippling anxiety.

4. Humility pays off during oral exams

During my second year I faced the usual oral exam to test whether I had sufficient background knowledge in my field before focusing almost exclusively on my thesis research. It was two hours long, precluding too many topics from being broached, and since I knew which professors were on the three-person committee I could reasonably narrow down the questions I might be asked. Even so, orals are one of the occasions in graduate school when they have you by the short hairs: the examiners, if so inclined, could ask just about anything they like. You do best by answering them as you would a police officer who has pulled you over for speeding. If you are prone to cockiness, it is better to affect nervousness, as even mild confidence may inspire some to take you down a peg, forcing you to repeat part of the exam some months later.

5. Avoid advanced courses that waste your time

After my three-person thesis committee was formed, I met with it to discuss advanced technique courses that I might need. Given that I now worked with phages multiplying in bacterial cells, I was in no position to tell the new resident whiz in bacterial physiology, Irwin (Gunny) Gunsalus, that I could get by without learning to measure key bacterial metabolic enzyme reactions. He devised a short lab course for me exclusively, and I got the A customary in such circumstances. But when the genial plant geneticist Ralph Cleland suggested that I follow up his uninspired cytology course with his fall offering on histological methods, I bluntly declared the course a waste of my time. Always polite in the Hoosier way, Cleland looked pained but did not challenge me. Returning with me to his bacteriology lab afterward, Luria let me have it and warned I must never again show contempt to a faculty member. Gunny took my side, making my day, saying that I had shown the kind of intellectual directness ascribed to the young J. Robert Oppenheimer.

6. Don't choose your initial thesis objective

At the time of my first experiments, I was too naive to devise an appropriate first research objective or even to choose wisely among alternatives presented by others. I therefore started my research on a problem that interested Luria. He showed an immediate interest in my results and saw that the control experiments were done. Luria had, moreover, wisely started me off on a problem not crucial to the progress of his own research, and so at no time would I feel pressed for my experiments to keep pace with his agenda. Happily, my first experiments yielded positive answers and Luria was gracious enough not to add his name to the abstract I submitted to the Genetics Society summarizing my first months' findings. Soon I was deciding my own experimental course and I would change my direction several times over the next two years. The two journal papers that summarized my thesis results would likewise appear under my name only, even though Luria helped me greatly, effectively rewriting many of my sentences before they were shown to other committee members, making the articles much more readable. Despite this needed help, Luria allowed me to feel that the papers belonged to me for better or worse and that I was working on behalf of no one but myself.

7. Keep your intellectual curiosity much broader than your thesis objective

Once a thesis is under way, it can feel like an all-consuming marathon. But my graduate experience was much enhanced by excellent courses I took over most of the time I was working on my thesis. There was always an alternative stimulation when my experiments weren't yielding the desired results. My favorite courses required long term papers and made me read original papers on topics I never would have delved into otherwise. Particularly influential in my intellectual development was the long paper I wrote at Tracy Sonneborn's suggestion on the German biochemical geneticist Franz Moewus's controversial experiments using the green algae *Chlamydomonas*. A recent course on scientific German let me read his original papers, including some published during the war and not generally known. Though Moewus's veracity had been challenged on the basis of results that seemed statistically too perfect, Sonneborn was intuitively persuaded by Moewus's elegant demonstrations of how genes control enzymatic reactions. Believing I had found new ways to interpret his data, I, like Sonneborn, also wanted to believe in Moewus's results. Afterward, Tracy incorporated part of my term paper analysis in a long review of

Moewus's work, but to our mutual dismay Moewus was found several years later to have faked his data. It was not a pleasant outcome, but if nothing else it was a valuable lesson about the dangers of wishful thinking in research, one better learned poring over someone else's work rather than one's own.

4. MANNERS FOLLOWED BY THE PHAGE GROUP

I **REACHED** New York City in mid-June 1948 after an overnight ride from Chicago on the Pennsylvania Railroad. At McKim, Mead, and White's Beaux Arts masterpiece Penn Station, I carried my luggage to an adjacent Long Island Railroad platform for the hourlong trip on to Cold Spring Harbor. A taxi whose base was the small wooden train station then brought me to the head of the inner harbor on whose innermost western shores lay Cold Spring Harbor Laboratory. I was let off in front of Blackford Hall, the summer center for the Lab where everyone ate, and whose upstairs dormitory space contained seventeen austere, concrete-sided single rooms. In one of these I was to stay all summer. Downstairs, in addition to the dining room, there was a lounge with a fireplace, a large blackboard, and three imposingly baronial wooden chairs that had been there ever since Blackford's construction in 1906.

Then the laboratory was effectively divided into two parts: a year-round Department of Genetics, supported by the Carnegie Institution of Washington, and the Biological Laboratory, a largely summertime effort under the patronage of the wealthy local estate owners. The latter body organized the summer courses and prestigious June meeting, the Cold Spring Harbor Symposium on Quantitative Biology, as well as providing housing and lab benches for summer visitors such as Luria and Delbrück. In 1941, the Yugoslav-born and Cornell-trained Milislav Demerec, the new director of the Department of Genetics, also took control of the community-supported Biological Laboratory, changing its emphasis from physiology and natural history to the study of the gene, his own interest. During his first year as director, Demerec staged the 1941 Cold Spring Harbor Symposium on Genes and Chromosomes. To this seminal meeting came both Hermann J. Muller and Sewall Wright, as well as Max Delbrück and Salvador Luria, for whom Demerec provided space to work together on phages in the Jones Laboratory.

After America entered World War II later that year, Demerec deployed most Biological Lab space to war-related activities. Jones Laboratory, however, being unheated, remained unoccupied and was thus still available to Delbrück and Luria by the time summer came. There were effectively no other wartime summer visitors except for the German-born ornithologist Ernst Mayr, then based at the American Museum of Natural History in New York City, whose research on evolution was complementary to Demerec's interest in genetics. With the director's natural affinity for European-born scientists, Cold Spring Harbor's atmosphere rapidly changed from that of a Yankee bastion, with a history of research in eugenics and an unmistakable anti-Semitic bias, to an international institution whose vigor much depended on foreign visitors, the matter of whose being in some cases Jewish posed no impediment to either scientific involvement or social acceptance.

Cold Spring Harbor was a place where reasoning through numbers was paramount long before Demerec's rule. In 1930 the biophysicist Hugo Fricke was appointed to the Biological Laboratory's staff, and 1933 saw the first Cold Spring Harbor Symposium on Quantitative Biology, its main purpose being to bring physicists and chemists together with biologists to help decipher the molecular basis of life phenomena. It was in this sense quite natural that physicists such as Delbrück and later Szilard were so warmly welcomed into the Cold Spring Harbor community.

My first afternoon's inspection of the digs and labs for the summer visitors revealed much physical dilapidation, the general atmosphere being that of a run-down summer camp. In fact, some visitors lived in tented cabins sited on the grass behind Blackford Hall, itself centrally heated only many years later. Both Hooper House and Williams

Canvas tents behind Blackford Hall sheltered some summering scientists.

*The Laboratory towed in the Cold Spring Harbor
Firehouse to house summer scientists in 1930.*

House, then used to house summer scientists with families, were built in the 1830s as "tenements" for workers in the antebellum whaling industry. Equally rundown was the three-story Firehouse, whose name dated from its original use as the town's first fire station before the Biological Lab bought it in 1930 for $50 and transported it across the harbor on a barge to provide more summer housing.

The research buildings housing the Department of Genetics dated from the first decade of its existence and, while of far sturdier construction, had an unmistakable turn-of-the-century smell. The main laboratory, an Italianate 1904 building, had a library on the first floor with labs on the top floor and in the basement. On two sides it was surrounded by the cornfields of the Brooklyn-raised and Cornell-trained geneticist Barbara McClintock, whom Demerec had recruited in 1941. Before coming to the symposium that June, she had resigned from the University of Missouri, where her sharp, independent mind was not so well appreciated. Demerec, however, with his eye for real talent, soon got Carnegie's permission to offer her a modest salary and second-floor space in the 1912 Animal House. There mice strains predisposed to cancer had been under intensive study since the early 1920s.

Despite the extraordinarily beautiful surrounding harbor and hills, the state of the labs and residences gave first-time visitors the distinct impression that Cold Spring Harbor would not likely long remain a site of high-powered science. But I then had no concern for the state of the buildings as long as they had the facilities needed for my phage experiments. Our IU contingent was to work in the 1927 Colonial Revival–style Nichols Building, which housed the Biological Laboratory's only two scientists, Vernon Bryson and Albert Kelner. Both did experiments on mutagenic agents in bacteria, and we could use their steam sterilizer (autoclave) and oven to prevent unwanted contamination of our bacterial cultures.

Several days later, Luria arrived with his New York–born wife, Zella, a Ph.D. student in psychology whom he'd met soon after arriving at IU. She was expecting their child at the end of the summer and welcomed the prospect of Blackford meals over preparing food in their Williams House apartment, furnished with castoffs from local estate

Tracy M. Sonneborn and Barbara McClintock at Cold Spring Harbor in 1946

owners. Renato Dulbecco by then also was on hand, having driven east in a secondhand Pontiac that he subsequently would use to take his wife and two young children back to Bloomington after their arrival from Italy. One of its first uses on the East Coast was to drive Max Delbrück and his wife, Manny, to the Marine Terminal at La Guardia Airport, where they boarded a Pan Am flying boat to England. From there they were to go on to Germany, where Max's family had suffered badly in the war, his brother, Justus, dying of diphtheria in a Russian prison camp after earlier being incarcerated by the Nazis.

Back for their sixth straight summer were Ernst Mayr, his wife, Gretel, and their two almost teenage daughters. He had long been one of my heroes, not only for his ornithological expeditions in the late 1920s to New Guinea and the Solomon Islands, but even more because of his seminal 1942 book, *Systematics and the Origin of Species.* I had excitedly read it during my last year at the University of Chicago, along with the equally influential *Genetics and the Origin of Species* by Theodosius Dobzhansky, a professor at Columbia University and frequent visitor to Cold Spring Harbor to see his close friend Demerec.

In our Nichols Lab room, Dulbecco and I soon began daily experiments to see whether UV light and X-rays inactivate phages by causing localized damage to one or more of their genes. Luria had arrived hoping that the phenomenon he observed, called multiplicity reactivation (whereby UV-damaged phage particles could somehow still multiply), was best explained by the independent replication of undamaged genetic determinants. But Renato's experiments soon began to show that the genes surviving a given UV dose replicated more slowly than their unirradiated counterparts. Conceivably there were no independently multiplying sets of phage genes. Instead they might linearly arrange on one or several chromosomes. If so, multiplicity reactivation was the result of crossing-over events between phage chromosomes, with genetic determinants from different particles coming together as a complete set of undamaged phage genes.

The summer session mood suddenly transformed when the three-week phage course started on June 28. Attendance in the dining hall almost doubled. It was the fourth year the course was given, and the first time without Delbrück. In his place the main instructor became Mark Adams from New York University's School of Medicine, who had taken the course two years earlier and been converted to full-time phage research. The fourteen students included the medically trained Bernard Davis, working on tuberculosis at Cornell Medical School; Seymour Benzer, a physicist from Purdue University; and Gunther Stent, a newly minted polymer chemistry Ph.D. from the University of Illinois, who was to start phage research with Delbrück in the fall at Caltech. I already knew much of what was presented, with the exception of the very new results of August (Gus) Doermann. Prior to coming to Cold Spring Harbor as a junior staff member, Doermann had learned to work with phages as a postdoc under Max at Vanderbilt. Gus described new experiments that for the first time could reveal the number of progeny phages present in infected bacteria as a function of the time following infection. Most exciting, he found that after attachments to their host bacteria, infecting phage particles become transformed into noninfectious replicating bodies. But there was no way then to know what these "replicating forms" were at the chemical level.

Gus and his southern-born wife, Harriet, lived in Urey Cottage, a tiny wooden house built during the Depression to employ idle carpenters. It was named after Harold Urey, the famed Columbia University chemist whose discovery of heavy water won him the Nobel Prize in 1934. Harriet's genteel background had not prepared her for the manners of someone like Delbrück, who could brutally criticize others at seminars and then afterward chat amiably with them at meals. Nor could she fathom how Manny Delbrück could leave her one-year-old son, Jonathan, with a babysitter to travel around Europe. Even more incomprehensible was Luria's support for Henry Wallace's new left-wing Independent Progressive Party, whose nominating convention in Philadelphia he had just attended with such enthusiasm. Upon his return, Luria organized a Saturday evening Wallace "corn party" in Jones Lab to raise money for the third-party ticket. Virtually everyone in the lab went, irrespective of their politics, to eat clams steamed in its autoclave and to drink beer. This was too much, however, for Harriet, who with Gus and his lab assistant, Maryda Swanstrom, picketed the party bearing large signs with the message "Wallace for President, Luria for Vice."

Always the favorite object of gossip was the frugality of the heavyset Milislav Demerec, whose verbal response to most propositions, good or bad, was "Do." Often as dusk fell, we would spot him roaming the lab grounds turning off lights in buildings whose occupants had gone home for the night. Getting him to agree to major repairs except those absolutely necessary was virtually impossible. Hilariously, he once refused to repair the cracked toilet seat in the Williams House flat of Leonor Michaelis, the great German-born biochemist, who was quite used to getting his way in his luxurious New York City Rockefeller Institute.

Soon after becoming director, Demerec moved into Airslie, the early-eighteenth-century wooden farmhouse at the northern end of the grounds that until 1942 had been part of a grand estate of more than one hundred acres, belonging to Henry deForest, whose main house, Nethermuir, dated from the 1850s. The estate included a large stable as well as a very beautiful English-style garden that won many accolades for the Olmsted brothers, who designed it. Early in the century, deForest's talents as a lawyer representing J. P. Morgan's railroads helped him multiply an already hefty family inheritance. As a director of the Southern Pacific Railroad, he had his own private rail coach. But by the time he died, in 1938, time and the Depression had taken a toll and his assets, once exceeding $70 million, ultimately were reduced to only $8 million. Also soon gone was the once grand Nethermuir, said to have mysteriously burned down on a winter's night in 1945 after

Mrs. deForest had moved permanently to her New York City apartment on Park Avenue.

With air-conditioning nonexistent outside a few lab rooms, the best way to beat Cold Spring Harbor's summer heat was to go swimming at high tide off the dock in front of Jones. When time permitted, we would swim off the sand spit, the body of land about a half mile long that almost closed off the inner harbor from the outer. At the Jones dock, there was a lab canoe that we used to paddle past the posh Moorings Restaurant on the eastern shore on our way to the village library, or the ice cream parlor, only three minutes away, which served super hot-fudge sundaes. There was also a tub-like boat owned by Sophie, the rotund blond fourteen-year-old daughter of the geneticist Theodosius Dobzhansky. She was spending the summer helping Barbara McClintock cultivate her cornfield. Frequently she saw reason to join the baseball games in the field to the east of the Carnegie main lab, fielding balls that careened toward McClintock's prized corn.

The Lab did not then own much of the land on the way to the sand spit. A large parcel still belonged to Rosalie Gardiner Jones, who lived in a mansion about a mile up the road to the Syosset train station. Born in 1885, the product of marriage between local estate-owning families, Rosalie had a long career as an activist suffragette and lawyer. Early in a life of more than ninety-five years, she was married off in a high-society wedding to Clarence Dill, a United States senator from Washington. But it was a brief union, as Rosalie soon found that she was not treated as an equal partner and the senator was dismayed to find his wife an untidy sort who buried garbage on the grounds of their home. Later Rosalie's most dependable companions were her white tennis shoes and the herd of goats that accompanied her on inspection tours of her various properties that dotted the shores of Cold Spring Harbor.

I first saw Rosalie when she drove her goat-bearing truck to the ramshackle mid-nineteenth-century wooden house she owned just north of the Firehouse. Without warning the Lab, she had rented it to migrant Jamaican farmworkers who had come to Long Island to help harvest vegetables on nearby farms. Quickly Demerec went to the police, telling them that the house, which we called Rosy's Cozy, was not fit for human inhabitants. Within days he obtained a court injunction ordering her tenants removed. Rosalie responded by bringing in the NAACP and claiming that the migrant workers would not have been asked to leave if they had not been "colored." But as soon as the NAACP representative saw the state of Rosy's Cozy, he too concluded that no one should live in such filth.

On Long Island some four hundred grand landed properties existed at the peak of the Gold Coast era in the 1920s. To better see what they had been like, one Friday night I walked north several miles along the western shore of Cold Spring Harbor. After passing a large abandoned boathouse from those roaring years, I was soon at the base of the 150-foot Cooper's Bluff, which looked out over Center Island to the Connecticut shore six miles away. Across the half-mile channel that led into Oyster Bay, the sounds of dance music drifted from the posh Seawanhaka Corinthian Yacht Club, a world I would never enter that summer. A paved estate road led up from the shore to a village road that took me to the Cove Neck police booth. There I was given directions for the two-and-a-half-mile walk back to the Lab. Later I was to learn I had been trespassing on the property of the children of Theodore Roosevelt, whose nearby summer home, Sagamore Hill, was then still inhabited by his second wife.

There was much tension in the air whenever Luria talked science with the slightly younger biochemist Seymour Cohen, who lived with his wife in an adjacent Williams House flat. Having taken the phage course two years earlier, Seymour had subsequently switched from studying typhus to investigating phage replication as a chemical phenomenon. He had come for the summer from the University of Pennsylvania to do experiments in Gus Doermann's lab measuring the time course of phage DNA synthesis during phage multiplication. While his wife and Luria's could be civil to each other, Seymour sensed all too clearly that, as a chemist, he would never be more than an outsider among the Delbrück-led phage group.

The phage course ended in late July with the juvenile antics of its graduation ceremony, the first without Delbrück, who had not yet returned from Europe. Draped in a white sheet over my shorts, I was given the role of his ghost. I was almost as tall as Max and wanted to

*I played his ghost in 1948, but when Max Delbrück was present
he took care of the phage graduation party himself. Here he is, standing,
with a flotation device around his neck and his hand on a bottle of beer.*

move about like him, befitting the godlike way his spirit dominated the occasion. Beer was an integral part of the evening, and the resulting mayhem provided the perfect pretext for younger scientists to become better acquainted with the female Carnegie assistants, individuals of both groups doubtless knowing that life would get lonelier once the summer had ended. The course had brought together Gunther Stent and Nao Okuda, a liaison that lasted all summer despite one unfortunate rendezvous in a field full of poison ivy. I spent the later part of the evening playing ping-pong under the Blackford porch with Sophie Dobzhansky or Nancy Collins, whose thick-rimmed glasses reflected the disdain she showed for girls who primped but had nothing going on between their ears. Nancy, a product of Vassar, was out for the summer from New York University to assist her thesis adviser, Mark Adams, with the phage course. She knew all too well that most men had only short-term objectives; like me, she was biding her time until the right person came along.

Frequently on the tennis court in front of Jones Lab was the physical anthropologist William H. Sheldon hitting with his much, much younger third wife. Sheldon, connected to the College of Physicians and Surgeons of Columbia University, was putting finishing touches on his about-to-be-published book *Variations of Delinquent Youth.* He also was at work on his newest opus, an *Atlas of Men,* and he gave a Thursday evening lecture about it in late July. He presented his thesis that human bodies should be viewed as composites of three qualities, ectomorphy (linearness), mesomorphy (muscle), and endomorphy (roundness), each derived from a different embryonic germ layer. The proportion of different tissue types dictated not only somatotype (kind of body) but also individual temperament. The lecture, profusely illustrated with photos of nude men, might easily have been regarded as a front for deviant sexuality but for the laughable extremes that predominated: torturously thin ectomorphs as well as an equal number of blubbery endomorphs. By Sheldon's scheme, which assigned numerical values to indicate proportion of tissue types, the ideal bodies were designated 3-4-3 physiques, derived from all three germ layers with a modest excess of muscles. Luria was not in the audience, not wanting to dignify Sheldon's obvious crap. Afterward Sheldon let me know I was a 6-1-1, an ectomorph having only minimal muscles to accompany my skin and nervous system. He did not mention the part of his theory holding that, overall, ectomorphs were relatively likelier to wind up in mental institutions.

Equally off the wall was the talk the following Thursday by Richard Roberts, who drove up from Washington, where he was a senior member of the Department of Terrestrial Magnetism of the Carnegie Institution. To our dismay, he had walked into Blackford with a golf bag, evidently prepared to spend much of his visit on the nearby links. A Princeton-trained nuclear physicist, in 1939 he made measurements on the number of neutrons released during uranium fission, data that were crucial to the development of the atom bomb. For this vital war research, he later received the Medal of Merit from President Truman. Wanting afterward to move into biology, Roberts had taken the 1947 phage course, whose students also included Leo Szilard and Philip Morrison, a former student of Robert Oppenheimer, who had done theoretical calculations for the first atom bomb. To everyone's surprise

Roberts's first foray into biology owed nothing to the phage course of the previous year, since his subject was extrasensory perception. Though Luria again bluntly declared to everyone that he would not dignify junk, curiosity got the better of virtually everyone else at Cold Spring Harbor. The lecture focused on the already highly controversial experiments at Duke University of parapsychologist J. B. Rhine. Soon Roberts was pulling cards out of carefully shuffled decks and predicting their values. Either he or the audience clearly had gone bonkers. At the end of the talk, the Berlin-born insect geneticist Ernst Caspari, drawing on his impeccable Continental politeness, thanked Roberts for his very unique talk.

The Delbrücks were then back from Germany, and had taken up residence in the flat on the north side of Hooper House. The lounge at Blackford soon became the Lab's intellectual center, its blackboard covered with thermodynamic equations involving alternative steady states. When in Germany, Max's interest was drawn to the topic through conversations with the noted physical chemist Karl Friedrich Bonhoeffer, his close friend. Karl's brother Klaus had been married to Max's sister, Emmi, before he was killed together with his other brother, Dietrich, a theologian, by the Nazis in 1945. In trying to understand how signals are propagated along nerve fibers, Bonhoeffer had developed equations that Max suspected to be applicable to the sudden changes in surface antigens on paramecia that Tracy Sonneborn was then keenly investigating at IU. Max did not like Sonneborn's own explanation, which postulated heredity determinants in the cytoplasm. Often working with Max to see if Bonhoeffer's equations were actually valid was Gunther Stent, free to do as he wished now that the phage course was over. Renato Dulbecco was also brought into action, revealing mathematical abilities not ordinarily associated with an M.D. I would silently watch them bat about equations, knowing I was not up to their mathematical manipulations.

At the end of the phage course, our IU group moved across Bungtown Road to the classroom lab in the Davenport Building that Gunther Stent was also making use of for the rest of the summer. My experiments were now even more closely tracking Renato's, studying progeny phages produced in bacteria simultaneously infected with genetically distinguished active and irradiated inactive phages. Happily, I found that genes from phage parents rendered inactive by X-rays, like those from UV-killed phages, were incorporated into the progeny phages. But whether this gene transfer resulted from a crossing-over process or through an independent assortment of independently replicating genes, as Luria still believed, remained unclear.

In mid-August, my parents took the train east for a two-day visit. Dad much liked being with Ernst Mayr and exchanging memories of great birds each had seen. Perhaps that inspired him to splurge and give me a glimpse inside the Moorings Restaurant, whose lofty prices effectively placed it out of bounds for the Lab's scientists. To drink beer or eat clams on the half shell, we normally went farther along the shore to the dependably seedy Neptune's Cave, whose open clam bar on Harbor Road attracted day-trippers out from the city. From Cold Spring Harbor, my parents went on to New Haven, where my father's physicist brother Bill and his family lived near Yale, and where my sister, Betty, was living for the summer.

After the Lurias returned to Indiana, Renato drove Max, Manny, and me to Cape Cod to spend several days at the Marine Biological Lab in Woods Hole. On our way back, I was dropped off at New Haven, where my sister was working at the Winchester Repeating Arms factory. That summer Betty had come from the University of Chicago to be part of a Students in Industry program, under the auspices of the Yale Divinity School, that brought together some ten socially conscious college girls with an equal number of male Chinese students belonging to the Chinese Christian Student Organization. All went well at the large Hillside Avenue residence belonging to the Divinity School until the Chinese students learned they were making rifles destined for the "corrupt" Chiang Kai-Shek in his losing fight against the advancing communist armies. They decided they could not in good faith continue working for the enemy, though they had no qualms about letting the American girls make more rifles to obtain funds needed for the group to remain in New Haven for the rest of the summer.

The week following, Renato and I closed down our Davenport lab, returning our pipettes, test tubes, and water baths to the storeroom for storage until the next summer. Renato's wife and two children were about to arrive by boat from Italy, and soon they would be driving to Bloomington. After a much awaited late August lecture by Ernst Mayr on the "species problem," the summer effectively came to an end. Of the summer crowd, only the Delbrücks and myself stayed on, taking our meals in the Carnegie dorm basement dining room, which opened up to serve the unmarried residents living in the rooms above. Max and Manny were in no hurry to be back in Pasadena since its hottest days normally occur in the early fall. I had the lucky excuse of needing to stay in the East until mid-September, when I was to meet Luria in Washington at the annual meeting of the Genetic Society. Earlier in the summer, Luria had greatly raised my morale by asking me to give a brief talk concerning my data on X-ray-killed phages to complement the paper he would be giving on his and Renato's latest data.

On her way back to Chicago from New Haven, my sister came to Cold Spring Harbor to see how I was living. I soon brought her to the Delbrück flat to meet Max and Manny and see how the phage group operated. She quickly sensed the tight knit of our club, where phages were at center stage and there was little sympathy for those scientists who did boring, if not stupid, experiments on other organisms. My hero worship for Max surprised Betty since his directness and self-confidence were antithetical to my parents' gentle empathy for those who could not make it to the top. She, moreover, was not at all prepared for Manny's free spirit and her obvious attraction to talented men.

The day before I took the train down to Washington, I walked south on the Cold Spring Harbor path bordering the mill ponds with the objective of reaching Oheka, the banker Otto Kahn's grand folly, whose immense château-like outline could be seen from the sand spit. It lay beyond the Cold Spring Harbor railway station, from which a half-mile-long private road led across the grounds of an equally private eighteen-hole golf course to the many-turreted edifice, whose more than one hundred rooms made it the second largest private dwelling in the United States upon its completion in 1917. To enhance the views from its yet-to-be-built windows, Kahn had brought in countless truckloads of sand upon which to set Oheka higher than other buildings on Long Island. His name was once legendary, with the financial power of the Kuhn-Loeb investment house that he ran with Jacob Schiff rivaling that of the House of Morgan. The Duke of Windsor spent a night in his grand home during his visit to America in the early 1920s. For more than twenty-five years, until his death in 1933, Kahn single-handedly covered the deficit of the Metropolitan Opera. But upon his death at the height of the Depression, the grand Oheka could find no master of sufficient means and was effectively abandoned until it was brought into service for war-related research.

Now it was again vacant and unattended when I found an open door to the garden and wandered amid the cracking plaster of its once grand rooms, from whose windows I could make out the distant Connecticut shore. Walking back more slowly so I could munch upon the wild blueberries thriving on the underlying sandy soil, I mused whether Oheka would ever come back to life. Basically I didn't care since its world was one I had no need to enter. In Cold Spring Harbor the world of the gene already had its waterfront country club, admittedly without the eighteen holes of golf. And though we knew no imperious grandes dames to pay us visits, we had our own equivalent in eccentric theoretical physicists now manic about genes as well as atoms. And instead of seeking instructions from the skies, we had Max's beat to follow.

Remembered Lessons

1. Use first names as soon as possible

From our first meeting, Max Delbrück called me Jim and likewise wanted everyone to call him Max. Among the phage group gathered around him and Salva Luria, no one was given a professorial designation unless in jest or when someone's apparent pomposity needed to be put down. Titles, like neckties, imply differences in rank or age but science moves best when all are treated as equals.

2. Banal thoughts necessarily also
dominate clever minds

I used to frequently position myself at meals near Max Delbrück, hoping to profit from sharp dissections of new experiments or criticisms of badly thought out ideas. On some days, conversation sparkled, particularly when a visitor brought new facts or gossip about friends from his European past. More often, however, Max found it more compelling to discuss a student's new girlfriend or who had beaten whom in tennis that afternoon. I was discovering that most high-powered minds do not daily generate new ideas. Their brains mostly lie idle until the input of one or more new facts stimulates their neurons to resolve the conundrums that stump them.

3. Work on Sundays

A fixed sabbath from experiments does not jibe with the reality of the human brain. It rests effectively only when it does not want to work and is satisfied with what it has done. With few exceptions, the time frame of experiments cannot be predicted, and mental hibernation should not be preassigned to a regular day on the calendar. An unanswered experimental question is bound to remain in your consciousness. Work done on weekends, in fact, can be more fun than that done on weekdays. You would not be there unless your experiments were going well.

4. Exercise exorcises intellectual blahs

Experiments or ideas should drive you forward but never should be counted on to keep you on an even emotional keel. Success is gratifying and failure is not, but failure is a necessary feature of the work: if your experiments work all the time or your ideas never stop coming, you likely are aiming at goals not worth pursuing. To counter the ups and downs in neurotransmitter levels that are a natural part of a career such as science, incorporate plenty of physical exertion to get outside your head regularly. Following Max Delbrück's example, I began running several times daily to and from the sand spit. Tennis, however, was my favorite nonscientific pastime, particularly when a good player gave me a match that made me work. Then I felt good even though I lost most games. The relaxation that comes from strenuous exercise most likely reflects the physical-stress-mediated release of β-endorphins, the opiate-like human molecules whose expression is evolution's way of ensuring that humans engage in tasks that promote our long-term well-being.

5. Late summer experiments go against human nature

During the euphoria that comes with long June days, both hard work and hard play are possible. A full day of experiments in no way precludes early evening softball or volleyball games. But by early August, darkness creeps up on mealtime and yellow leaves begin to hint that fall is not that far away. So with the outside water temperature still rising to its early September highs, afternoon beach excursions make more sense than experiments easily put off to the next morning. The last weeks of August are usually best suited for vacations to distant places attractive enough that thoughts of science will fade no more than two or three days after arrival. Several-week vacations never hurt if you can afford them. And on beach walks toward the end of your vacation, your brain may even be sufficiently refreshed to mull over potential experiments you can undertake when back on home ground.

5. MANNERS PASSED ON TO AN ASPIRING YOUNG SCIENTIST

Upon my return to the less intense intellectual atmosphere of IU in the fall of 1948, I began following up Luria's observations from 1941 that phages suspended in simple salt solutions are much more sensitive to inactivation by X-rays than those suspended in nutrient-rich beef broth solutions. Unclear was whether phages indirectly killed by exposure to reactive molecules generated by X-rays striking surrounding water molecules possessed novel properties not found in phages killed by "direct" X-ray hits. Luria's earlier inactivation curves suggested that several indirect hits were required to kill a phage. In contrast, direct killing was long thought to result from a single ionization event.

While enjoying the first experiments of my own devising, I began anticipating the intellectual excitement that was to come from the impending mid-October weekend visit of Leo Szilard. Just turned fifty, Szilard was then a professor of biophysics and sociology at the University of Chicago, and was driven down by his much younger collaborator, Aaron Novick, also a participant in the 1947 Cold Spring Harbor phage course. Leo had recently received a small Rockefeller Foundation grant to support midwestern genetics meetings of his choosing. The barely five-foot-six Szilard invariably wore a tie with his suit, never trying to hide the potbelly that reflected his fondness for food and aversion to exercise.

Born in Budapest in 1898 to prosperous parents, the extraordinarily intelligent Leo became a physicist in Berlin, where he knew Albert

Einstein well and taught modern physics between 1925 and 1932 with Erwin Schrödinger. As a Jew, he had the good sense to flee Berlin the month Hitler assumed power. Soon he was in England, where the fast flow of his ideas was not so well suited to the more stately flow of English science. He seldom spent more than a few months in any one location, and so there never seemed to be enough time for his theoretical hunches to be experimentally tested. Moreover, his desire to seek patents for ideas that had commercial application made his English academic hosts think he valued money more than ideas. Here they were 100 percent wrong. It was only thanks to money from his German patents, one with Einstein, that Leo could afford to stay in science.

No one in England, moreover, knew of the personal anguish attending his 1933 revelation in London that a nuclear disintegration releasing more neutrons than it consumed would unleash the great energy of the atom described by Einstein's famous $E = mc^2$ equation. If the technique for creating such fission events were to fall into the hands of the Nazis, allowing them to build atomic bombs, they would have all the power they needed to conquer the world. Secretly Leo assigned his patent to the British Admiralty, revealing it to close friends only after the uranium atom was experimentally split in Berlin in 1939. Until then Leo had incorrectly targeted first beryllium and then indium as elements likely to produce the necessary chain reaction.

Immediately Leo tried to stop his physicist friends outside the Third Reich from publishing more on uranium fission. But that cat was let out of the bag when, against Leo's advice, Frédéric Joliot in Paris soon published his findings that uranium-235 fission generates two neutrons, not one. Leo then became obsessed with seeing that the United States moved ahead as fast as possible toward the construction of atomic weapons. It was he who first composed Einstein's famous fall 1939 letter to Franklin Roosevelt and, a year later, co-opted Enrico Fermi, the 1938 Italian Nobel Prize winner, by then a refugee at the Columbia University Physics Department, to work on uranium fission. Two years later, they moved from Columbia to the University of Chicago, where their nuclear reactor first went critical in early December 1942. Judged too independent to be part of any military-led team,

Leo, unlike Fermi, was kept from the subsequent bomb-making activities at Los Alamos by General Leslie Groves, then in charge of the Manhattan Project. But as soon as the first bombs went off, Leo worked incessantly to see that civilians—not the military—were in control of the Atomic Energy Commission.

Now Leo had set his sights on cracking the genetic basis of life. After taking the 1947 phage course, he saw the need for frequent assemblies of bright people to inform him of new facts to chew on. That Bloomington weekend, however, provided no take-home lesson, either from my brief presentation on X-ray-killed phages or from Renato's much more sophisticated experiments on UV-inactivated phages. The most important new results presented, in fact, came from Szilard and Novick themselves. Over the past six months they had become convinced that despite Max Delbrück's very public reservations, Joshua Lederberg's 1946 demonstration of genetic recombination in *E. coli* was correct. Gleefully Leo wrote both Max Delbrück and Salva Luria that he would eat his hat if someone was able to disprove his and Aaron's new experiments. In fact, they were soon to find out that Lederberg had already published similar confirmatory data.

After Szilard and Novick went back to Chicago, Renato returned to experiments where he began seeing irreproducibility, a problem never before encountered in Luria's lab. Agar-coated plates expected to show statistically equivalent numbers of multiplying phages often yielded wildly disparate counts. Then, on a mid-November afternoon, he noticed the agar plates on the top of piles had more phage plaques. Plates lower in the piles, less exposed to the recently installed fluorescent lights, had fewer plaques. This observation was confirmed the next day, telling Renato that visible light reverses much UV damage, an effect soon called photoreactivation. Immediately I tested whether photoreactivation occurred with X-ray-damaged phages but was disappointed to discover only a small, possibly insignificant effect. Salva, then at Yale for a week of lectures, only learned of the light bombshell when Renato and I met him just before Thanksgiving at Szilard's second get-together at the University of Chicago. Immediately Salva feared that his past multiplicity reactivation results might have been badly compromised by inadvertent light exposure. But Renato put his mind at ease, pointing out that Salva already had reproduced multiplicity reactivation under light conditions insufficient for photoreactivation.

In turn, Salva reminded Renato of a letter from Cold Spring Harbor, from Albert Kelner, which had arrived in Bloomington just before he left for Yale. In it Kelner excitedly told Luria of his discovery early in September that UV-killed bacteria and fungi could be resurrected by visible light. For many preceding months Kelner had also been plagued by irreproducible results that he thought might be due to variations in the temperatures to which his UV-exposed bacterial cultures were exposed. Just after Dulbecco and I left Cold Spring Harbor, Kelner found that light, not temperature, was the uncontrolled variable messing up his experiments. Luria did not show Kelner's letter to Dulbecco, only casually mentioning the result to him, and Dulbecco made no connection to his own irreproducible results.

We were then all gathered in front of a blackboard in Szilard and Novick's lab, located in the former synagogue of an abandoned Jewish orphanage in a run-down neighborhood adjacent to the University of Chicago. As a physicist, Leo knew that visible light alone was unlikely to furnish sufficient energy to reverse UV damage. But he was intrigued to learn from Renato that visible light had no effect on free phages. It only worked after the damaged phages had entered their host bacteria cells. Immediately Leo began to speculate whether UV-induced mutations would also be reversed by visible light under such circumstances. To answer this question, he and Novick did experiments over the next six months that showed UV-created mutations were "cured" by visible light in the same proportions that visible light reactivated UV-killed bacteria.

Though I was also finding some of my "indirect effect" experiments difficult to reproduce, Chicago was not the place to say so. It was only just before Christmas that I realized that my IU X-rays were producing not only very short-lived free radical molecules but also much more stable peroxide-like intermediates that persisted after the X-ray machine was turned off. Not anticipating this, I had not been controlling the time from X-ray exposure that I did my assays for viable phages. My experiments continued, more confusing than enlightening, until our next Szilard-inspired get-together in Bloomington, just prior to a meeting at Oak Ridge National Laboratory at which Luria and Delbrück were to speak.

Initially Luria had wanted me to make a presentation as well, believing that my results indicated that what had been described as direct X-ray effects were actually caused not by ionizations directly breaking vital phage components but by the effects of reactive chemicals such as free radicals generated by X-rays within the phage particles. Szilard, however, sitting in the front row of our Bloomington gathering, unsentimentally tore that argument apart. He focused on my observation that purified phage particles suspended in nonprotective media lost their ability to kill bacteria every time they were inactivated. All too clearly I saw that I must do more experiments before I ventured again to speak even informally.

My bungled presentation was then followed by a slapstick exchange between Szilard and Novick. Novick was to present their seemingly paradoxical data produced following mixed infection of bacteria by the closely related phages T2 and T4. Sensing that no one followed Novick's argument, Szilard stood up to compound the confusion. Unlike me, however, they truly had something important to say in explaining results that had baffled Delbrück three years earlier. In the end Max had to clarify what they jointly failed to get across. Following mixed infection by T2 and T4, some progeny particles had the T2 genotype but the T4 phenotype, and vice versa. In fact, Leo and Aaron had pulled off some neat science. At the time, Max wrongly thought it elegant but not very important, and he urged Leo and Aaron not to publish their results. Only two years later did they write them up for *Science*.

Photoreactivation discussions dominated the Oak Ridge meeting. Albert Kelner talked about results that he had rushed to publish upon learning that Renato Dulbecco also had found photoreactivation. Renato, believing that his discovery had been made effectively independent of Kelner's, initially did not refer to him in a short note later prepared for publication in *Nature*. Upon then reading Dulbecco's proposed phage photoreactivation manuscript, Kelner felt robbed. In his eyes, Dulbecco must have been influenced by his prior work as reported in his letter to Luria. Immediately responding to Kelner's unhappiness, Renato revised his *Nature* note to cite prior knowledge of Kelner's observation.

As soon as I got back to Bloomington, I felt I had to re-convince Luria that I could do meaningful science. So I stopped irradiating impure phage solutions capable of generating peroxides and instead focused on the biological properties of purified phages killed by short-lived free radicals. Soon I had irrefutable evidence that they truly differed from phages killed directly by X-rays. Not only were several damaging events needed to inactivate them, but when so killed, they were incapable of multiplicity reactivation.

By then I was eagerly anticipating going to Caltech for the summer. The phage group would have gone back to Cold Spring Harbor except for Manny's expecting the second Delbrück child in August. Her need to be in Pasadena provided the perfect excuse for a summer in California. Renato's trip, however, was to be one-way since Max had just induced him to move there with the promise of greater intellectual independence and stability than he now had at Indiana. In the meantime, I was finishing up assisting in the bird course, knowing by then where to lead field trips toward the crow-sized pileated woodpecker. Because of its more southerly range I had never been able to see one around Chicago.

The day-and-a-half train trip to California was largely sleepless, and through the train's windows I began to spot magpies and lark buntings as the cornfields gave way to prairie land. I was more than groggy upon my arrival at the Athenaeum, Caltech's faculty club. Its upstairs loggia housed a row of camp-like cots, one of which was to be my cheap berth for the summer. Upon dropping off a rucksack filled with all my possessions, I made the short walk to the Kerckhoff Laboratory, built twenty years before to house biologists brought together by T. H. Morgan, who came to Caltech in 1928. Morgan had been dead now four years, and the new head of the Biology Division, George Beadle, had been brought down from Stanford to bring Caltech into the era of the genetics of microorganisms.

One of Beadle's first moves was to entice Max to move back to Caltech. Beginning late in 1946, he and Manny lived only ten minutes away by foot, in a new one-story ranch-style home they built on one of the few remaining vacant lots near Caltech. When I first went there for supper, I was much impressed by the large main room with a fireplace graced by a large painting done by Jeanne Mammen, a Berlin friend of Max's from the 1930s. Before Hitler's rise to power, she drew and painted the demimonde, but such art would have been degenerate according to Nazi orthodoxy, and the painting now dominating the Delbrück sitting room drew inspiration from Picasso's classical canvases of the 1920s. Much less memorable was Manny's food. She was not one to pore over cookbooks while Max was back at the lab for seminars. Mexican-spiced ground meat and lots of avocados satisfied her and Max, eating being more a practical necessity than a pleasure for them. They cared more about quality in conversation, chamber music, and tennis partners, and were thrilled by the smells and sights of the California outdoors.

Salva would not be arriving for another two weeks, and I wanted to greet him with new experiments on phages killed by hydrogen peroxide. Studying it in Bloomington was never high on the agenda Luria had set for me; my few such experiments were done virtually by stealth. Tantalizingly they hinted that peroxide-killed phages had biological properties identical to those inactivated by X-ray-irradiated bacterial lysates. If so, there would be good reason to believe that organic peroxides were the phage-killing molecules present in my irradiated phage lysates. Working then on the lab bench next to me was Gunther Stent, already a year in the Delbrück lab, studying how tryptophan influenced the attachment of phage T4 to *E. coli* cells. Also there was the French scientist Elie Wollman, whose Jewish parents, scientists themselves, had perished in the Nazi camps. Wollman never felt at ease with the young German chemist Wolf Weidel, who cohabited with him in their laboratory room. But Gunther, though also Jewish, soon became good friends with Wolf, whose Teutonic upbringing made it painful for him to call Max by his first name.

Getting reproducible survival curves took more time than I anticipated, and the Lurias arrived before I had results to show Salva. Subsequent nonstop lab orgies, during which I was in the lab long past midnight, alternated with manic weekend car trips instigated by the indefatigable Carleton Gajdusek, who had completed his degree at Harvard Medical School two years before and now was supposedly getting postdoctoral experience in both Max's lab and the chemist John Kirkwood's. My first such camping trip ended when the corrugated road gave out five hours below Ensenada in Baja California. Two weekends later, we embarked on an even more insane nonstop drive to Guaymas on the Gulf of California. There for the first time I saw huge man-of-war birds circling over the harbor. A primitive ferry ride across the Río Yaqui interrupted our journey onward to Ciudad Obregón, where 110-degree temperatures finally persuaded Carleton that you could die from the heat. On subsequent weekends, Carleton's extreme traveling turned toward the much cooler Sierras, where on one occasion the rest of our party reached the summit of Mt. Whitney long after he had gone on and descended into a valley to the west.

Such weekends away kept my morale high long after I'd reached the inescapable conclusion that Pasadena was strictly for retirees. Indeed, the average age of the residents in Caltech's hometown was higher than that of any other American city of note. Even on the Caltech campus it was hard to detect a pulse outside the labs and libraries. Social life was most accurately described as nonexistent. Mindful of this reality, Gunther Stent had moved into a canyon house above Caltech occupied by several European postdocs. In this way, he entered into the orbit of the younger chemists associated with Linus Pauling. Late that summer Linus, after virtually bumping into me at the Athenaeum, gave me a big grin. Initially I assumed that Max and Linus must have interacted often, since when Max first arrived he and Linus coauthored a short note to *Science* attacking the notion that putative like-with-like attractive forces would play a role in copying genetic information. More recently Max had become wary of Pauling's self-aggrandizement, though he always remained alert to reports of what Linus was up to from his postdocs.

Of all the phage crowd gathered there, I was most at ease with the Doermanns. The high point of my summer came in late August when, on the Athenaeum courts, I took two tennis sets from Gus. In the evenings we would often go into central Pasadena to a restaurant where we had earlier spotted two striking blondes about my age. They, however, never reappeared, nor did our two-hour-long drive to the Pacific Coast beach next to Caltech's marine station at Corona del Mar prove more fruitful for girl-gazing. But at least by then I had accomplished my summer lab objective of showing that peroxide-treated phages had biological properties identical to those killed by X-ray-irradiated phage lysates.

I was thus prepared to speak several days later before an afternoon phage group meeting presided over by Max. The week before, we had listened to the young physicist Aage Bohr talk about the philosophical implications of quantum uncertainties. Here he was a surrogate for his father, Niels, who first had mesmerized Max in the early 1930s. Besides Max, only Gunther pressed Aage for more precise information about his father's supposed philosophical insights. In my back-row seat, I understood not a word of either Aage's thrust or Max and Gunther's counterarguments. In contrast, my talk about three types of X-ray-killed phages revealed no grand paradoxes, nor was much brainpower needed to understand my conclusions. Remembering acutely my April debacle in front of Szilard, I stuck to facts and was careful not to imply any form of breakthrough for radiation biology—much less toward understanding the gene.

The next day in his office, Max told me not to despair of my unexciting results. Instead I should consider myself lucky not to be in Renato's shoes, forced into an emotionally consuming photoreactivation rat race irrelevant to the much more important question of how genetic information is copied. Now was the time for me to concentrate on learning to do science as opposed to winning an experimental race whose outcome would surely be only marginally significant a decade later. George Beadle also reassured me that I was not off course. To my surprise, he had popped in to hear my seminar and, soon afterward, invited me to dinner at his modest home nearby. Like Max, he was no longer doing experiments, instead getting his scientific kicks from walking about the Kerchoff labs to see what the younger graduate students and postdocs were up to. Already he was justly famous for work at Stanford using the mold *Neurospora* to find genes coding for metabolic pathway enzymes. At forty-five, he didn't see himself making another such conceptual advance.

On the lookout for
girls in Corona del
Mar, California

In early August the Berkeley bacteriologist Roger Stanier gave a seminar on bacterial metabolism. Roger was still a bachelor, and his presence led to the arrival several days later of the Hopkins Marine Station graduate student Barbara Wright. Failing to attract Roger's notice, she caught the eye of Wolf Weidel, who asked her to join him, Gunther Stent, and a Biology Department secretary for a camping weekend on Catalina Island. After Gunther's date vamoosed in favor of a reconciliation with her husband, I was asked to go along out of pity for my being otherwise condemned to another weekend of Pasadena desolation. All went well until the four of us got off the boat at Avalon, the only town, and learned that camping was forbidden. Believing it a ruse to make us rent hotel rooms, we walked toward the island's opposite side hoping to find there a secluded beach on which to roll out our sleeping bags.

On an increasingly blistering afternoon, we realized too late that only goats had ever walked our path snaking down a cliff face to the ocean several hundred feet below. Neither Gunther nor Wolf initially wanted to seem cowardly in front of Barbara, while I awkwardly declared I was going back alone. But after a few more steps downward, the others agreed to turn back. Then, without warning, Gunther's backpack, momentarily off his shoulders, rolled down the steep incline to the beach below. Faced with the prospect of spending real money to replace the bag and its contents, Gunther and Wolf again inched downward, reaching the ocean some twenty minutes later. Soon, however, they found it impossible to retrace their steps. After an hour passed with them out of sight searching for alternative upward paths, Barbara and I saw no option but to go back to town.

It was already dusk as we went back along the route we had taken, our bare legs constantly assaulted by spines from the prickly pears that, along with the goats, were the island's principal inhabitants. In town, I anticipated renting rooms so we could shower. But to save money, Barbara insisted that we go back to just beyond the outskirts, where we found a large vacant field to plop down our sleeping bags. There at dawn we were arrested for camping out on the golf course. Later, back at the police station, by saying we were pelican-seeking biologists, I got the police chief to help mount an apparently futile rescue mission for Gunther and Wolf above the cliffs in his Jeep. Returning empty-handed to town, we soon happily spotted Gunther and Wolf near the boat dock. After sunrise they had found a chimney-like indentation in the cliff face that let them squirm upward until they reached a spot from which they could scramble to safety. They were still shaking, knowing they had put their lives at great risk. By then I had lost my reading glasses. Gunther was even more annoyed that neither Barbara nor I had spotted the expensive camera he'd left behind in his pursuit of his backpack. And so no pictures of our weekend misadventure survive.

Soon after my early September return to Bloomington, Luria asked me to give a bacteriology seminar in which I talked about Seymour

Cohen's experiments at Penn showing that phage-infected bacteria synthesized no bacteria-specific molecules, but instead phage-specific DNA and protein. How to go beyond these neat results of Seymour's was not at all clear. No chemist had yet mastered the basic chemistry of either proteins or of the two nucleic acids DNA and RNA. Even Linus Pauling remained then mostly in the dark. Though with great antici-pation I went to IU's chemistry auditorium to hear him give the fall Sigma Xi lecture, his talk was about the structure of antibodies as opposed to that of the gene.

I wanted to move on as a postdoc to a lab where I could learn nucleic acid chemistry. But no obvious place suggested itself during a late October evening meal with Salva and Zella. Resolution did not come until just before Christmas, during the second of that fall's Szilard-sponsored Chicago get-togethers. By then Joshua Lederberg was part of our in-group, with his first appearance given over to a four-hour monologue on perplexing bacterial genetic results from his University of Wisconsin lab. To the second gathering also came the biochemist Herman Kalckar, now back in his native Denmark after spending the war years mainly in St. Louis. A participant in Max's first phage course, Herman professed the desire to use some of his rare, recently synthesized radioactive adenine to study phage replication. So both Max and Salva quickly urged me to move on to Kalckar's lab, located in Copenhagen, not far from Niels Bohr's institute and the intellectual tradition that had spawned Max's first interest in biology. Happily, Kalckar instantly said he would accept me, and I promptly applied for postdoctoral fellowships that would allow me to move to Copenhagen.

At the same time I was repeating many previous key experiments of my thesis to reassure Salva that its conclusions, though not earth-shattering, were at least solid. This task was over by the end of Febru-ary, allowing me to complete a first draft for my thesis before I flew to New York in mid-March to be seen by the selection committee for National Research Council postdoctoral fellowships. Though the bumpy flight made me awfully airsick, the interview went well and in less than two weeks I was awarded a prestigious two-year Merck fel-lowship. I had expected my coming summer to be spent in Oak Ridge with Gus Doerman, who had recently moved to the big Atomic Energy Commission biology lab there. But in early May, Gus told me his attempt to get me a security clearance had failed: my association with the left-wing Luria made me a risk. In the summer of 1948, a Cold Spring Harbor–sited FBI informant had attended the Wallace-for-president fund-raising corn party in Jones Lab to which virtually all the Cold Spring Harbor community, myself included, not so earnestly went. Max came to my rescue, asking me back to Caltech for June and July before I joined him in Cold Spring Harbor for the August phage meeting. By then, Salva had virtually rewritten my thesis, making my late May thesis exam mainly perfunctory.

Only in my last year at Indiana did I have a real girlfriend. She was a perky, dark-haired fellow graduate student in the Zoology Department, Marion Drasher. In early December, I took her to a local production of J. B. Priestley's play *An Inspector Calls*. Soon I was intensely in love, particularly after Christmas of 1949, when we were in New York City together with several other Bloomington students for the big annual AAAS meeting. At the beginning she was the reluctant one, citing her several years' advantage in age. Our relative roles slowly reversed upon our return to Bloomington, however, with me increasingly resistant to making long-term plans together. I was after all anticipating my trip to Copenhagen within six months, and in no sense wanted to be tied down. How to go back to just being friends eluded both of us, and when we parted in June I felt bad about being so emotionally inconsistent.

Much of my second Caltech interval I spent converting my thesis into the first of two manuscripts for the *Journal of Bacteriology*. For a few days, I did experiments with a T5 mutant with a lengthened life cycle, but Max chided me that I was wasting my time in the absence of a defined experimental objective. So instead of hanging around the lab without real purpose, I was more frequently in the library or on the Athenaeum tennis court. For several days I was with George Beadle at Caltech's marine biology station, to which he had gone to collect invertebrate specimens. Then Renato and I climbed Mt. San Jacinto again, going through clouds to reach its treeless top, almost twelve thousand feet above Palm Springs. Several days later, my mood suddenly turned serious with the start of the Korean War. But when I passed through Chicago on my way to Cold Spring Harbor, and then by boat to Copenhagen, my draft board offered no objection to my going abroad as long as I kept them informed of my address.

At the Cold Spring Harbor phage meeting in late August, Salva was at ease about the setback to his multiplicity reactivation theory, no longer believing such experiments held vital clues about phage genes. His morale was again high, thanks to a new observation of the frequency of spontaneous mutants among individual bacteria, which he believed showed that genes duplicated by a process akin to binary fission. In contrast, Max still wanted to pull sense out of multiplicity reactivation curves, interpreting Renato's latest examples to suggest the possibility of two forms of DNA—one genetic, the other nongenetic. If phages were indeed so constructed, this might explain Lloyd Kosloff and Frank Putnam's finding at the University of Chicago that when DNA was tagged by introducing radioactive isotopes, only half the DNA of infecting phage particles is transferred to their progeny particles. Here Seymour Cohen pointed out that these radioactive progeny would only have their label in genetic DNA and would in turn pass 100 percent of their labeled DNA to second-generation progeny particles.

My mind turned again to potential second-generation experiments as soon as the seasickness-inducing vessel *Stockholm* docked in Copenhagen. There I found Kalckar keen that I focus instead on enzymes that make the nucleoside precursors of DNA. But after a week listening to Herman's almost indecipherable English, I saw that experiments with nucleosides would never get at the essence of DNA. I, however, could not figure out a graceful way to tell Herman that my time was better spent going back to phage experiments. Deciding to say nothing, I was soon cycling each day through the center of Copenhagen to the State Serum Institute, where Herman's friend Ole Maaløe was keen to follow up the private phage course given to him by Max at Caltech.

Long before we began producing second-generation results,

Kalckar's marriage suddenly collapsed. No longer enzyme-driven, Herman was obsessing about Barbara Wright, the feminine component of our calamitous camping trip to Catalina Island the year before. Like me, she was a new postdoc in Kalckar's lab, as was Gunther Stent, who'd come from Caltech the month before. Delusionally believing Barbara's Ph.D. thesis had earth-shattering implications, Herman hastily arranged an afternoon get-together at the Institute for Theoretical Physics, where Gunther and I listened to her explain her experiments to Niels Bohr. Herman then proudly acted as intermediary between Barbara, his putatively visionary biologist, and Bohr, the inarguably visionary physicist. After an hour passed, Bohr politely excused himself.

By winter's end, Ole and I finished our experiments, getting the answer that the first-generation progeny transmitted DNA to their second-generation progeny no better than the parental particles. No evidence suggested the existence of two forms of DNA. Though this was not the answer we had hoped for, Max thought it sufficiently important to submit the resulting manuscript to the *Proceedings of the National Academy.* Soon Herman himself felt the need to absent himself from his lab, announcing that he and Barbara would spend April and May at the Zoological Station in Naples. Maintaining the facade that I was still his postdoc, Herman asked me whether I wanted to join him in learning more about the marine biology that Barbara had been raised on. Instantly I accepted, for I had no potentially exciting phage experiment on the horizon.

Just before I left Copenhagen, there was a small microbial genetics gathering to which came the Italian aristocrat Niccolò Visconti di Modrone, whose keen intelligence I had first witnessed the preceding August at Cold Spring Harbor. Just back in Milan from Caltech, Niccolò said I must stop off in his ancestral city to hear a performance at La Scala. Upon meeting my train from Copenhagen, he noticed that my rucksack held all my belongings, and deduced I was without a dark suit. So he arranged for us to go to the same Weber opera but on different nights. At the genetics department in the nearby small university town of Pavia, Niccolò and I bumped into Ernst Mayr, whom Niccolò also knew from Cold Spring Harbor. After we all visited the

At the State Serum Institute in Copenhagen, 1951. Gunther
Stent is on the far left, Ole Maaløe is third from left, Niels
Jerne is standing, and I am sitting in front of Niels.

ancient Certosa di Pavia, we had supper in the large farmhouse of Niccolò's equally tall and good-looking brother, just back from China.

I would have considered such acculturation alone ample justification for my spending two months in Italy, but a small, high-level meeting on macromolecular structure in the Zoological Station auditorium provided an even better excuse. Until that mid-May gathering in Naples, I had assumed no one would soon understand the detailed, three-dimensional structure of DNA at the atomic level. Since genetic information, which was encoded within DNA, varied, each different DNA molecule most likely presented a different structure to solve. But my pessimism, born of chemical naivete, lifted dramatically after a talk by the youngish King's College London physicist Maurice Wilkins. Instead of revealing disorganized DNA molecules, DNA in his X-ray diffraction pictures was yielding patterns consistent with crystalline assemblies. Later he told me that the DNA structure might not be that difficult to solve since it was a polymeric molecule made up from only four different building blocks. If he was right, the essence of the gene would emerge not from the genetic approaches of the phage group but from the methodologies of the X-ray crystallographer.

Despite my obvious excitement at his results, Maurice did not seem to judge me a useful future collaborator. So upon arriving back in Copenhagen, I wrote Salva seeking help in finding another biologically oriented crystallographic lab in which I could learn the basic methodologies of the structural chemist. Salva delivered after a meeting in Ann Arbor at which he met the Cambridge University protein crystallographer John Kendrew. Then just thirty-four, John was seeking an even younger scientist to join him. With Salva having spoken well of my abilities, he agreed to my coming aboard to learn crystallographic methodologies from him and his colleagues at the recently established Medical Research Council (MRC) Unit for the Study of Structure of Biological Systems.

By then I was again studying the transmission of radioactive labels from parental to progeny phages, knowing that early in September Max Delbrück was coming to Copenhagen for an international poliomyelitis conference. When his ship arrived, Gunther, Ole, and I went to the Copenhagen dock to greet Max with a large poster saying "Velkommen Max Mendelian Mater." The congress itself was a routine affair except for dinner at Niels Bohr's home within the Carlsberg Brewery. Its founder had long before arranged that his opulent domicile should always be occupied by Denmark's preeminent citizen. Luckily, I was not seated near Bohr, who was likely to be expressing thoughts that no one around him, Danish or foreign, could understand.

Soon I was in England to meet John Kendrew's coworker Max Perutz, to make preparations for my coming to Cambridge in early October. Though John was still in the States, my meeting with Perutz and his boss, the Cavendish Professor of Physics, Sir Lawrence Bragg, went well and I took that night's train to Edinburgh for a two-day peek at the Scottish Highlands near Oban. In returning by train to London, I was engrossed in Evelyn Waugh's *Brideshead Revisited*.

*Max Delbrück arrives
in Copenhagen,
September 1951. From
left: Gunther Stent,
Ole Maaløe, Carsten
Bresch, and Jim Watson*

Delbrück was ending his European trip with visits to André Lwoff
and Jacques Monod at the Institut Pasteur, and so from London I flew
to Paris. There on a Sunday afternoon, after watching Monod nimbly
scale the big boulders in the woods at nearby Fontainebleau, I said
goodbye to Max as he boarded a plane at Orly. Though Max was
highly skeptical of my foray into a Pauling-like structural chemistry,
he did not choose this occasion to say so. Instead he wished me well
and I felt the creeping apprehension of knowing that I would no
longer be part of the world in which grace and the fall from it could be
comfortably predicted by asking, "What will Max say?" Soon I would
be somewhere he did not matter.

Remembered Lessons

1. Have a big objective that makes you feel special

No one within the phage group of 1950 would have denied our air of self-importance or our sense of being a happy few. The disciples of George Beadle and Ed Tatum working with *Neurospora* on gene-enzyme connections never came together with such esprit de corps. Max Delbrück's personality was a big factor. His reverence for deep truths and commitment to sharing them unselfishly was saint-like. But these virtues attend many uninspired minds as well and were never the key to the fervor of his acolytes. Instead it was his great commission that we go to the heart of the gene, in search of its genetic and molecular essences. To obssess over less fundamental goals made no sense to Max. Phages, being virtually naked genes that yielded answers after only a night's sleep, had to be the best biological tools for moving forward fast. Legions of graduate students across biology were pursuing things worth knowing but perhaps not worth devoting one's life to. The quest for such an unrivaled prize of indisputable significance fired in our imaginations a devotion such as religion fires in others', but without the irrationality.

2. Sit in the front row when a seminar's title intrigues you

By far the best way to profit from seminars that interest you is to sit in the front row. Not being bored, you do not risk the embarrassment of falling asleep in front of everybody's eyes. If you cannot follow the speaker's train of thought from where you are, you are in a good place to interrupt. Chances are you are not alone in being lost and most everyone in the audience will silently applaud. Your prodding may in fact reveal whether the speaker indeed has a take-home message or has simply deluded himself into believing he does. Waiting until a seminar is over to ask questions is pathologically polite. You will probably forget where you got lost and start questioning results you actually understood.

Now, if you have suspicions that a seminar will bore you but are not sure enough to risk skipping it, sit in the back row. There a dull, glazed expression will not be conspicuous, and if you walk out, your departure may be thought temporary and compelled by the call of nature. Szilard did not follow this advice, habitually sitting in a front row and getting up abruptly in the middle of talks when he'd had too much of too little. Those outside his close circle of friends were relieved when his inherent restlessness made him move on to a potentially more exciting domicile.

3. Irreproducible results can be blessings in disguise

A desired result in science is gratifying, but there is no contentment until you have repeated your experiments several times and got the same answer. Al Hershey called such moments of satisfaction "Hershey heaven." Just the opposite feeling of maddening inferno comes from irreproducible results. Albert Kelner and Renato Dulbecco felt it before they found that visible light can reverse much UV damage. Delbrück, struck by how long this phenomenon remained undiscovered, put it down to fastidiousness. He described what he called "the principle of limited sloppiness." If you are too sloppy, of course you never get reproducible results. But if you are just a little sloppy, you have a good chance of introducing an unsuspected variable and possibly nailing down an important new phenomenon. In contrast, always doing an experiment in precisely the same way limits you to exploring conditions that you already suspect might influence your experimental results. Before the Kelner-Dulbecco observations, no one had cause to suspect that under any conditions visible light could reverse the effects of UV irradiation. Great inspirations are often accidents.

4. Always have an audience for your experiments

Before starting an experiment, be sure others are interested in the premise. Mindless minor variations on prior good science will generate yawns in the world beyond your lab. Though such almost repetitive motions are good ways for students to learn lab techniques, they should be seen as exercises and not as real science, with their results publishable only in journals that hotshot scientists never read. Now, trying to break new ground may lead to consequences that seem worse than yawns, and you must be prepared for most of your peers to think you are out of your mind. If, however, you cannot think of at least one and preferably several bright individuals who can take appreciative notice of what you are doing, your tenacity may very well indicate that you are either stupid or crazy.

5. Avoid boring people

Social gatherings of even successful academics are no different from gatherings of any professional cohort. The truly interesting are inevitably a small subset of any group. Don't be surprised when arriving at some senior colleague's house for dinner if you feel an unexplainable desire to leave when you learn whom you're seated next to. Routinely reading the *New York Times* at breakfast will expose you to many more facts and ideas than you are ever likely to acquire during evenings with individuals who in most instances haven't had to think differently since getting tenure. Unless you have reason to anticipate a very good meal or the presence of a fetching face, take care not to accept outright any invitations to senior faculty's homes. Leave open the possibility that a sixteen-hour experiment might keep you from coming. If you later find out that someone you want to meet will be there, make known your sudden availability and come gallantly with a small box of chocolates to enjoy with the coffee.

6. Science is highly social

In high school there is a domain of facts and ideas in which you can succeed separate from the world of hanging out with your peers. Once you get into science, however, worlds collide, and not only your fun but also your professional success demands you know as much about your peers' personality quirks as you do about their experiments. Gos-

sip is a fact of life also among scientists, and if you are out of the loop
of what's new you are working with one hand tied behind your back.
The intellectual vitality of the phage group drew not only from its
meetings but also from constant visits to one another's labs, often for
joint experiments. Particularly at the start of your career, you should
seize any chance to see how other labs function and talk about results
that might be interpreted in new ways. It's all too natural when young
to see one's peers merely as competitors. Some of that is necessary and
appropriate, but scientific knowledge is not a zero-sum game: there is
always something more to be discovered, and getting to know your
colleagues can only help you get a piece of the prize.

7. Leave a research field before it bores you

When I decided to abandon the genetic approach of the phage group
in favor of learning X-ray crystallography to go after the three-
dimensional structure of DNA in Cambridge, I was in no way bored
with the work of Max Delbrück and Salva Luria. My last phage experiments in Copenhagen were still very rewarding. By then, however, I
was more and more drawn to finding the DNA structure, and meeting
Wilkins gave me good reason to believe the phage group, for all its
high purpose, was not the way. In science, as in other professions and
in personal involvements, individuals too often wait for abject misery
before effecting change that makes perfect sense. In fact, there is no
good reason ever to be on the downward slope of experience. Avoid it
and you'll still be enjoying life when you die.

6. MANNERS NEEDED FOR IMPORTANT SCIENCE

I **ARRIVED** in Cambridge in the fall of 1951 sensing a majesty of place and intellectual style unmatched anywhere in the world. Its great university, reflecting almost nine hundred years of English history, first centered itself along the banks of the river Cam, whose modest waters move northeast across East Anglia to the market city of Ely. There its massive twelfth-century cathedral had long towered over the vast flat fenland marshes that emptied into the Cam forty miles from the shallow waters of the Wash, over which tidal waters from the North Sea still roar twice daily. It was the draining of the fens over many centuries that created the rich agricultural fields and wealth of the great East Anglia estate owners. Their benefactions in return helped create along the "backs" of the Cam the many elegant student residences, dining halls, and chapels that already many centuries ago marked out Cambridge as a market city of extraordinary grace and beauty.

For most of its history, Cambridge University was highly decentralized, with the teaching exclusively carried out by its residential colleges, among which Trinity was long the grandest, having enjoyed the matchless patronage of Henry VIII. In a room off the Great Court had lived the young Newton, whose greatest science was done in his twenties and thirties before he went up to London to be master of the mint.

Until the mid-eighteenth century, the primary role of the colleges was to educate clergy for the Church of England, the mission carried out by fellows (dons) who were themselves required to remain unmarried while part of college life. Only in the nineteenth century did science become an important part of the Cambridge teaching scene. Charles Darwin's serious excitement about natural history and geology came from his exposure in the early 1830s to these disciplines at Christ's College. Over the next half century, the responsibility for instruction increasingly shifted away from the colleges to newly created academic departments under university control. In 1871, the Duke of Devonshire, Henry Cavendish, donated funds for the creation of the Cavendish Laboratory and the appointment as the first Cavendish Professor of James Clerk Maxwell, whose eponymous equations first unified the dynamics of electricity and magnetism. Upon Maxwell's early death at age forty-nine in 1879, the twenty-nine-year-old John William Strutt (Lord Rayleigh), famed for his ideas on optics, became the second Cavendish Professor of Physics. In 1904, he was to win a Nobel Prize, as would the next four successors to the chair: J. J. Thomson (1906), Ernest Rutherford (1908), William Lawrence Bragg (1915), and Nevill Mott (1977).

By the start of the twentieth century, Cambridge stood out as one of the world's leading centers for science, of the same rank as the best German universities—Heidelberg, Göttingen, Berlin, and Munich. Over the next fifty years, Cambridge would remain in that rarefied league, but Germany's place would be supplanted by the United States, much strengthened by its absorption of many of the better Jewish scientists forced to flee Hitler. England similarly much benefited from the arrival of some extraordinary Jewish intellectuals. If Max Perutz had not had the good sense to leave Austria in 1936 as a young chemist, there would have then been no reason for my now moving along the Cam.

Though winning the great struggle against Hitler had drained England financially, the country's intellectuals took pleasure in knowing that their country's great victory was much of their own making. Without the physicists who provided radar for British aviators during the Battle of Britain, or the Enigma code breakers of Bletchley Park who successfully pinpointed the German U-boats assaulting the Allies' Atlantic convoys, things might have turned out very differently.

Emboldened by the war to think boldly, the then tiny Medical Research Council (MRC) unit for the Study of Structure of Biological Systems was doing science in the early 1950s that most chemists and biologists thought ahead of its time. Using X-ray crystallography to establish the 3-D structure of proteins was likely to be orders of magnitude more difficult than solving the structures of small molecules such as penicillin. Proteins were daunting objectives, not only because of their size and irregularity but because the sequence of the amino acids along their polypeptide chains was still unknown. This obstacle, however, was soon likely to be overcome. The biochemist Fred Sanger, working less than half a mile away from Max Perutz and John Kendrew at the MRC lab, was far along the path to establish the amino acid sequences of the two insulin polypeptides. Others following in his steps would soon be working out the amino acid sequences of many other proteins.

Polypeptide chains within proteins were then thought to have a mixture of regularly folded helical and ribboned sections intermixed with irregularly arranged blocks of amino acids. Less than a year before, the nature of the putative helical folds was still not settled, with the Cambridge trio of Perutz, Kendrew, and Bragg hoping to find their way by building Tinkertoy-like, 3-D models of helically folded polypeptide chains. Unfortunately, they received a local chemist's bad advice about the conformation of the peptide bond and, in late 1950, published a paper soon shown to be incorrect. Within months they were upstaged by Caltech's Linus Pauling, then widely regarded as the world's best chemist. Through structural studies on dipeptides, Pauling inferred that peptide bonds have strictly planar configurations and, in April 1951, he revealed to much fanfare the stereochemically pleasing α-helix. Though Cambridge was momentarily stunned, Max Perutz quickly responded using a clever crystallographic insight to show that the chemically synthesized polypeptide, polybenzylglutamate, took up the α-helical conformation. Again the Cavendish group could view itself as a major player in protein crystallography.

The unit's resident theoretician was by then the physicist Francis Crick, who at thirty-five was two years younger than Max Perutz and one year older than John Kendrew. Francis was of middle-class, nonconformist, Midlands background, though his father's long-prosperous shoe factories in Northampton failed during the Great Depression of the 1930s. It was only with the help of a scholarship from Northampton Grammar School that Francis moved to the Mill Hill School in North London, where his father and uncle had gone. There he liked science but never pulled out the grades required for Oxford or Cambridge. Instead he studied physics at University College London, afterward staying on for a Ph.D. financially sponsored by his uncle Arthur, who after Mill Hill had chosen to open an antacid-dispensing pharmacy instead of joining the family shoe business.

Unlike Max and John, who came into science as chemists and now possessed Ph.D.'s, Francis's doctorate was not completed. He had done just two years of thesis research, winning a prize for his experimental apparatus to study the viscosity of water under high pressure and temperature, when the advent of the war moved him to the Admiralty. After joining the high-powered group set up to invent countermeasures against German magnetic mines, his boss, the Cavendish-trained nuclear physicist Harrie Massey, gave him in 1943 the challenge of combating the German navy's latest innovation. In great secrecy, their shipyards had under construction a new class of minesweepers (*Sperrbrechers*) whose bows were fitted with huge five-hundred-ton electromagnets designed to trigger magnetic mines lying a safe distance ahead. Crick came up with the clever idea that a specially designed insensitive mine would not explode until the *Sperrbrecher* passed directly over it. By the end of the war, more than a hundred *Sperrbrechers* were so sent to the bottom of the ocean.

After Harrie Massey left to lead the British uranium effort at Berkeley, the Cambridge mathematician Edward Collingwood became Francis's mentor. He saw Francis both as a friend and as an invaluable colleague, inviting him for weekends to his large Northumbrian home, Lilburn Tower, and taking him to Russia in early 1945 to help decipher the workings of a just-captured German acoustic torpedo.

After the war's end, Francis's new bosses did not need to be as forgiving of his loud, piercing laughter or of the distaste for conventional thinking that often inspired it. Though formally made a member of the civil service in mid-1946, Francis soon lost interest in military intelligence and wanted a bigger challenge. He saw in biology the greatest range of potential problems to engage his inquisitive mind.

Apprised of Francis's desire for a radical change of course, Harrie

Massey sent him to see the physicist Maurice Wilkins at King's College London, which had a new biophysics laboratory. After the war, while still in Berkeley, Massey had changed Wilkins's life by giving him a copy of Erwin Schrödinger's *What Is Life?* Its message that the secret of life lay in the gene was as compelling to Maurice as it had been to me, and he soon began to make his move into biophysics. He would join J. T. Randall at St. Andrews and then move with him to London. Immediately he and Francis became friends, with Maurice soon asking Randall to offer a job to Francis. Randall thought better of it, though, correctly seeing Francis as a mind he could not control. The Medical Research Council, mindful of Francis's high wartime repute, came to his rescue and funded his learning to work with cells at the Strangeways Laboratory on the outskirts of Cambridge.

His task during the next two years at Strangeways—observing how tiny magnets moved through the cytoplasm of cells—did not win Francis any kudos. At best it was busywork that gave him time to seek out more appropriate challenges. These at last came when he moved his MRC scholarship across Cambridge to Max Perutz's protein crystallographic unit. Though his new job was no better paid, it would let him work toward the Ph.D., by then a prerequisite for meaningful academic positions.

By the time I came to Cambridge, Francis's forte was increasingly seen to be crystallographic theory, though his early forays in the field had not been universally appreciated. At his July 1950 first group seminar, entitled "The Theory of Protein Crystallography," he came to the conclusion that the methodologies currently used by Perutz and Kendrew could never establish the three-dimensional structure of proteins—an admittedly impolitic assertion that caused Sir Lawrence Bragg to brand Crick a boat rocker. Much more harm came a year later when Bragg presented his newest brainchild and Francis told him how similar it was to one he himself had presented at a meeting six months earlier. After the infuriating implication of his being an idea snatcher, Sir Lawrence called Francis into his office to tell him that once his thesis was completed he would have no future at the Cavendish. Fortunately for me, and even more so for Francis, Cambridge was unlikely to grant him the degree for another eighteen to twenty-four months.

I was by then having lunch with Francis almost daily at a nearby pub, the Eagle, which during the war was favored by American airmen flying out of nearby airfields. Soon we would be upgraded from desks beside our lab benches to a largish office of our own next to the connected pair of smaller rooms used by Max and John. In this way, Francis's ever irrepressible laughter would less disturb the work habits of other unit members. At our first meeting, Francis had spoken of his much valued friend Maurice Wilkins, who, like him, had made a wartime marriage that soon disintegrated with peace. Because he was curious to know whether Maurice's crystallography had generated any new, perhaps sharper X-ray photos from DNA, Francis invited him for a Sunday dinner at the Green Door, the tiny apartment on top of a tobacconist's on Thompson Lane, across from St. John's College. Earlier occupied by Max Perutz and his wife, Gisela, it had been home to Francis and his second wife, Odile, since their marriage two years before, in August 1949.

At that meal, we learned of an unexpected complication to Maurice's pursuit of DNA. While he was on an extended winter visit to the United States, his boss, J. T. Randall, had recruited to the King's DNA effort the Cambridge-trained physical chemist Rosalind Franklin. For the past four years in Paris she had been using X-rays to investigate the properties of carbon. Rosalind understood from Randall's description of her responsibilities that X-ray analysis of DNA was to be her responsibility solely. This effectively blocked Maurice's further X-ray pursuit of his crystalline DNA. Though not formally trained as a crystallographer, Maurice had already mastered many procedures and had much to offer. But Rosalind didn't want a collaborator; all she wanted from Maurice was the help of his research student Raymond Gosling. Now, though out in the cold for two months, Maurice could not stop thinking about DNA. He believed his past X-ray pattern arose not from single polynucleotide chains but from helical assemblies of either two or three intertwined chains bonded to each other in a fashion as yet to be determined. With the DNA ball sadly no longer under his control, Maurice suggested that if Francis and I wanted to learn more we should go to King's in a month's time to hear Rosalind give a talk on November 21.

*Rosalind Franklin,
while in Paris,
serving afternoon
coffee in evaporating
dishes*

Before it was time to go to London, Francis had reason to feel good about his place in the Cavendish. He and the clever crystallographer Bill Cochran derived easy-to-use mathematical equations for how helical molecules diffract X-rays. Each of them, in fact, did so independently within twenty-four hours of being shown by Bragg a manuscript from Vladimir Vand in Glasgow, whose equations they immediately saw as only half-baked. Theirs was an important achievement, for Francis and Bill had given the world the equations that could predict the diffraction patterns of specific helical molecules. The next spring I was to deploy them to show that the protein subunits of tobacco mosaic virus are helically arranged.

Suddenly, the best way to reveal DNA's 3-D structure was to build molecular models using Cochran and Crick's equations. Until a year previously this approach had made no sense since the nature of the covalent bonds linking nucleotides to each other in DNA chains was unknown. But after work by Alex Todd's nearby research group in the chemical laboratory at Cambridge, it was clear that DNA's nucleotides are held together by 3′-to-5′ phosphodiester bonds. A focus on model building was a way to set oneself apart from the alternative approach of focusing on X-ray photograph details being pursued at King's College in London.

On the day of the lecture, Francis was unable to go down to London and I went alone, still oblivious to the difference between the crystallographic terms "asymmetric unit" and "unit cell." As a result, the next morning I mistakenly reported to Francis that Rosalind's DNA fibers contained very little water. My error only came to light a week later, when Rosalind and Maurice came up from London to look at a three-chain model that we had hastily constructed. It had DNA's sugar-phosphate backbone in the center with the bases facing outward. Upon seeing it, Rosalind immediately faulted its conception, saying the phosphate groups were located on the outside, not the inside of the molecule. Moreover, we had proposed DNA to be virtually dry whereas, in fact, it was highly hydrated. And we got the unmistakable impression that the King's group considered the pursuit of the DNA structure to be their property, not one to be shared with their fellow MRC unit in Cambridge. All too soon we learned that Sir Lawrence Bragg was of the same mind, when he told us to refrain from all subsequent DNA model-building activities. In stopping us Bragg was not motivated solely by a need to remain on good terms with another MRC-supported group. He wanted Francis to focus exclusively on research for his Ph.D. and be done with it.

This debacle, however, would not have occurred if Francis and I had started to think as if we were chemists. Even without the King's X-rays, there were sufficient clues in the chemical literature that should have led us to propose a double helix as the basic structure of DNA. From the start we should have restricted ourselves to models in which externally located sugar-phosphate backbones were held together by hydrogen bonds between centrally located bases. Strong physical chemical evidence for bases so held together had come from the postwar experiments of John Gulland. In 1946, his Nottingham lab showed that within native DNA molecules the bases are so arranged as to hinder them from exchanging hydrogen atoms. These data suggested widespread hydrogen bonding between DNA bases. This insight was widely available, published by Cambridge University Press in the 1947 Society for Experimental Biology symposium volume on nucleic acids.

Furthermore, given Linus Pauling and Max Delbrück's prewar proposal that the copying of genetic molecules would involve structures of complementary shape, Francis and I should have reasonably focused on two-chain rather than three-chain models. Thinking this way, each DNA base should hydrogen-bond exclusively to one with a molecule of complementary shape. In fact, experimental data pointing to this conclusion, too, already had been published, most coming from the lab of the Austrian-born chemist Erwin Chargaff in New York. Without understanding the significance, he reported that in DNA the amounts of the purine adenine were roughly equal to the amount of the pyrimidine thymine. Likewise, the amount of the second purine, guanine, was similar to the amount of the second pyrimidine, cytosine.

The exact shape of such base pairs would depend upon where the atoms available for hydrogen bonding were located on each base. In 1951, few chemists knew enough quantum mechanics to make such inferences. So that fall we should have sought advice from the several British chemists trained in this esoteric field. In retrospect, Alex Todd's lab, after determining the covalent linkages in DNA, should have moved on to determining what the molecule looks like in three dimensions. But in those days, even the best organic chemists thought such problems were better left to X-ray crystallographers. In turn, most X-ray diffraction experts felt the time had not yet arrived to tackle biological macromolecules. In a sense, then, the field was wide open.

Even after he found the α-helix, Linus Pauling remained only moderately attentive to DNA, never seriously believing then that it had a genetic role. Even so, when hearing of Maurice Wilkins's crystalline photo, he asked to have a look, being misinformed that Maurice himself was not seriously trying to determine the structure. As that was precisely what Maurice was up to, he quickly replied that he wanted more time to look over the photo before releasing it to others. Undeterred, Linus wrote directly to the King's boss, John Randall, but this approach was likewise unsuccessful. Linus lost the scent until a year later at a summer phage meeting outside Paris, where he first learned of the work recently completed at Cold Spring Harbor by Alfred

Alfred Hershey's group at Cold Spring Harbor, 1952:
Niccolò Visconti, Martha Chase, Al Hershey, Constance
Chadwick, Neville Symonds, June Dixon, and Alan Garen

Hershey and Martha Chase, showing phage genes were also made from DNA. The news convinced Linus he must go after the DNA structure despite his lack of high-quality DNA X-ray photos. His voyage back to the States could have been a fortuitous opportunity. Also on board the transatlantic boat was Erwin Chargaff, who like Pauling had come to Europe to attend that summer's International Biochemical Congress in Paris. But instead of learning about the equivalence of A with T and G with C, Linus took an instantaneous dislike to his shipmate and avoided him all across the Atlantic.

Preoccupied much of the fall of 1952 with the race against Francis Crick for the coiled coil structure of α-keratin, Pauling turned to DNA in earnest only in late November. Soon he was very much attracted to a DNA model in which three sugar-phosphate backbones coiled around each other. He was hung up on three chains because of the reported high density of DNA. At no time did he seriously consider a two-chain molecule. To hold the three chains together, he conceived of DNA as uncharged, forming hydrogen bonds between opposing phosphate groups. Soon satisfied that he had found the general structure for nucleic acids, he wrote to Alex Todd a week before Christmas, adding he was not bothered that his structure provided no clues as to how DNA functions in cells. That problem was for another day. At no time did he ever take into account Chargaff's base compositions, published more than a year before in several journals. The essential parameters for Linus that December were bond angles and length, not what DNA did biologically or how it behaved in solution. It was immediately evident that the atoms of his model were not fitting together as neatly as they did in the α-helix. Even his best structure was stereochemically shaky, with several central phosphate oxygens uncomfortably close to each other.

Fearing that someone in England might beat him to the punch with a similar model, Linus hastily submitted a manuscript for publication in the *Proceedings of the National Academy.* Then he triumphantly sent two manuscript copies to Cambridge—one to Bragg, the other to his son Peter. We were instantly engulfed in anxiety until we realized that Linus had used hydrogen atoms belonging to the phosphate groups to hydrogen-bond the three chains together. We knew at once his model must be wrong since DNA—an acid—normally releases all its hydrogen ions in solution, and so Francis and I rushed around Cambridge to see whether the local chemical hotshots also found Pauling's concept totally implausible. Quickly reassured by Alex Todd that Linus had indeed made a gigantic chemical goof, I went down almost immediately to London to show the manuscript to Maurice Wilkins and Rosalind Franklin, the latter preparing to move to J. D. Bernal's group in Birkbeck College, where she would no longer work on DNA.

Maurice was more than relieved to learn that Linus was so far off base. In contrast, Rosalind was annoyed at my showing her the manuscript, tartly telling me that she had no need to read about helices. In her mind, the crystalline DNA A-form structure was most certainly not helical. In fact, six months before, she had sent out invitations to a July "memorial service" to celebrate the death of the DNA helix. Here Maurice thought that Rosalind had been badly deluding herself, and to prove it he impulsively showed me an X-ray photo that the King's group had been keeping secret since Raymond Gosling took it more than nine months before. Originating from a more hydrated B-form DNA fiber, this picture displayed unequivocally the large cross-shaped diffraction pattern to be expected from a helical molecule. My jaw dropped, and I rushed back to Cambridge to tell everyone what I had learned. In my mind we should not wait a moment longer before commencing to build models. Someone was bound to tell Linus that his three-chain model was dead on arrival. Sir Lawrence Bragg instantly agreed, and with him finally behind us Francis and I soon were back playing with cutout shapes. By then I realized that DNA's density did not, as I originally thought, rule out two strands as opposed to three. It thus made sense for me to focus first on possible ways for two DNA chains to twist around each other.

In fact, Rosalind also should have been focusing on two-chain DNA models. More than a year before, she had carefully measured her X-ray diffraction patterns from crystalline A-form DNA, looking for possible molecular symmetries. Finding her data compatible with three possible chemical "space groups," she went up to Oxford to get advice from Dorothy Hodgkin, then England's premier crystallographer, justly famed for solving the problem of the structure of penicillin. As soon as Dorothy saw that Rosalind was considering space groups involving mirror symmetry, however, she sensed crystallographic callowness. Experienced crystallographers would never postulate mirror symmetry for a molecule made up exclusively of 2-deoxy-D-ribose. Instead, Dorothy believed, Rosalind should now have been considering only the implications of the third monoclinic space group (a rectangular prism of three unequal axes). Upset by Dorothy's sharp putdown of her crystallographic acumen, Rosalind left Oxford, never to return. If she had gone instead to Francis for help, she would have immediately learned that the C2 monoclinic space group suggested that DNA was a double helix with its chains running in opposite directions.

Francis learned of DNA's monoclinic space group only through reading a nonconfidential King's progress report sent to Max Perutz in mid-February. By then, through a new burst of model building, I had found that a sugar-phosphate backbone of 20 Å diameter optimally repeats every 34 Å, the repeat distance measured in B-form DNA. Francis now argued, in light of Rosalind's space group, that the two chains must run in opposite directions. But I didn't initially buy this assertion, not understanding the underlying crystallographic symmetry argument. Until I knew how the centrally located bases bonded to each other, I didn't want to worry about backbone directions. Then, unknown to me, my model building was being hindered by faulty textbook descriptions for the structures of guanine and thymine. Using such false configurations, I had become momentarily excited about a pairing scheme similar to that found in crystals of adenine.

That scheme, however, would have given a 17 Å repeat along the helical axis, not the 34 Å figure observed by Rosalind. Happily, the Caltech structural chemist Jerry Donohue, then spending his sabbatical year in Cambridge, set me on the right track by arguing that the guanine and thymine hydrogens should have keto rather than the textbook-ascribed enol configurations. Needing only a day to incorporate Jerry's reasoning, I changed the locations of the hydrogen atoms on my paper-cutout models of thymine and guanine. Almost instantly I found myself forming the A-T and G-C base pairs we now know to exist in DNA. Coming a half hour later into our office that Saturday morning, Francis took only a few minutes to conclude that symmetry of the base pairs demanded that the chains run in opposite directions. Rosalind's monoclinic space group was in a true sense a prediction of a model derived by Francis and me from purely stereochemical arguments. The double helix had to be correct. All that remained to be done was to build a backbone segment and measure its atomic coordinates to show that all the bond lengths and angles in our model agreed with those previously found in smaller molecules. This task, which for the first time in months took Francis away from his desk, took less than three days to complete. The double helix was ready to be let loose upon the world.

Breaking the news to Wilkins that we very likely had solved the DNA structure was bound to cause his heart to spasm. A day after we

Francis and I posing with our morning coffee in our office,
just after the publication of our manuscript in Nature

had verified appropriate coordinates for all the atoms, a letter from him arrived informing Francis that Rosalind was out of King's and that Maurice was about to resume work on DNA. Perhaps to soften the blow, John Kendrew, not Francis, called Maurice to report that Francis and I had a promising novel structure for DNA. Coming up the next day, Maurice instantly recognized the double helices' elegant simplicity and agreed that it was likely too good not to be true. Aware that we would not have found the DNA structure without knowledge of X-ray results from King's, Francis and I suggested to Maurice that his name also be on the manuscript we planned to send to *Nature*. Without hesitation he declined, possibly not knowing how to deal with Rosalind Franklin and Raymond Gosling's equally important contributions. The April 25, 1953, issue of *Nature*, besides containing the nine-hundred-word description of our model, included separate continuing contributions from the two warring DNA groups at

There was at the Cavendish today a great air of excitement for Perutz and two of the younger men in the department were anxious to show Sir Lawrence (and GRP) what they have been up to in the last week. They believe they have really got the structure of nucleic acid from a crystallographic rather than a chemical standpoint. Their clue came out of the beautiful X-ray diagrams produced in Randall's lab and some of the work which had meanwhile been going on at Cambridge. They are just putting the finishing touches on a huge model about six feet tall which shows that the molecule is made up of two helical chains running parallel to each other and repeating at a distance of 34 Å. The two coils run one above the other, at equal distances from the center, and centripetal to them are a series of direct linkages between pyrimidine and purine rings which, in turn, are liked to the helices. They have discussed their model with Wilkins of Randall's group and have written up a description of it as a note for NATURE (which would probably appear in early May) which Sir L. today authorizes them to send. They are particularly excited about the possibility of showing it to Pauling, who will be here this week, but GRP also persuades them to show it to Todd first (and to this Sir L. gives full approval). If this structure were the correct one it would present a very good skeleton on which to affix some of the modern theories of chromosome duplication. Sir L. has some small reservations about it all, but the usually cautious Perutz is quite enthusiastic.

The two other chaps who have been associated with Perutz on the nucleic acid structure are J. D. Watson and F. H. C. Crick. Watson was trained in genetics and biochemistry at the University of Chicago and got his PhD under Luria. He went to Copenhagen on a Merck fellowship and has been at the Cavendish since September 1951. At present is on a Polio Foundation fellowship and in September will return to the USA to work with Delbruck at Caltech. Crick is a

1 April 1953 (Cambridge) - continued

physicist trained in chemistry by Andrade at University College. He worked on Navy problems during the war, then in 1946 got an MRC studentship to work with Hughes at the Strangeways. He was too quantitatively minded for Hughes's type of work and came to the Cavendish four years ago. Will get his PhD in X-ray crystallography this summer and is then going to the USA to spend six months with Harker and six months with Pauling. Both young men are somewhat mad hatters who bubble over about their new structure in characteristic Cambridge style and it is hard to realize that one of them is an American. Having in mind the lot that one sees in Bernal's lab it is hard to comprehend why crystallography should show such a sex bias in selecting out good-looking and apparently very stable and well-adjusted young women. The two chaps here are certainly not lacking, however, in either enthusiasm or ability.

Excerpt from the diary of Gerard Roland Pomerat, of the Rockefeller Foundation, describing his visit to the Cavendish Lab, dated April 1, 1953

King's. Maurice was later to write that his refusal to publish jointly with the two of us was the biggest mistake of his life.

In every sense solving the double helix was a problem in chemistry. Alex Todd facetiously told me that Francis and I were good organic chemists, not wanting to admit that a major objective in chemistry had been solved by nonchemists. In reality, Francis and I would not have been first to see the structure if Todd's fellow chemists had not done botched jobs. Linus had all the keys to unlock the DNA structure but inexplicably didn't use them that fall of 1952. Rosalind Franklin would have seen the double helix first had she seen fit to enter the model-building race and been better able to interact with other scientists. If she had accepted rather than rejected Maurice as a collaborator, the two of them could not have failed to realize the significance of the monoclinic space group. Dorothy Hodgkin's Oxford put-down of Rosalind as a crystallographer would then not have been the fatal wound that in retrospect it now seems.

In contrast, Francis and I were far from being on our own. One flight up was the clever Bill Cochran, who put the Bessel functions of helical diffraction theory into Francis's working vocabulary, whence they entered mine. Even more important, Jerry Donohue's spartan desk was no more than twelve feet from mine and Francis's when his quantum chemistry expertise squelched my initial desire to build a double helix based on like-with-like base pairing (e.g., A-A and T-T). The Cavendish was then a magnet for minds that wanted to be challenged by others of equal power. In contrast, Linus Pauling's Caltech was a chemistry garden of mortals hovered over by a god who saw no need to assimilate the ideas and facts of others. If Linus had only spent a few days in Caltech's libraries perusing the literature on DNA that fall, he would most likely had to have hit upon the idea of base pairing and would now be celebrated for both the α-helix and the double helix.

Virtually everyone who came to our now even more cramped Cavendish office to see the large 3-D model made in early April was thrilled by its implications. Any doubt as to whether DNA and not protein was the genetic-information-bearing molecule suddenly vanished. The complementary nature of the base sequences on the opposing chains of the double helix had to be the physical counterpart of the Pauling-Delbrück theoretical postulation of gene copying through the creation of complementary intermediates. DNA double helices as they exist in nature must reflect single-stranded template chains hydrogen-bonded to their single-stranded products of complementary sequence. Two of the three big questions in molecular genetics, what is the structure of DNA and how is it copied, were suddenly resolved through the discovery of base-pair hydrogen bonding.

Still to be ascertained was how the information conveyed by the sequence of DNA's four bases (adenine, guanine, thymine, and cytosine) determines the order of the amino acids in the polypeptide products of individual genes. Since there were known to be twenty amino acids and only four DNA bases, groups of several bases must be used to specify, or code for, a single amino acid. I initially thought the language of DNA then would be best approached not through further work on the DNA structure but by work on the 3-D structure of its close chemical relative ribonucleic acid (RNA). My decision to move on from DNA to RNA reflected the observation, already several years old, that polypeptide (protein) chains are not assembled on DNA-containing chromosomes. Instead they are made in the cytoplasm on small RNA-containing particles called ribosomes. Even before we found the double helix, I postulated that the genetic information of DNA must be passed on to RNA chains of complementary sequences, which in turn function as the direct templates for polypeptide synthesis. Naively I then believed that amino acids bonded to specific cavities linearly located on the surfaces of the ribosome RNA components.

Three subsequent years of X-ray studies—the first two at Caltech and the last back with the "Unit" in Cambridge, England, in which I was joined by the Pauling- and Harvard Medical School–trained Alex Rich—frustratingly failed to generate a plausible 3-D structure for RNA. Though RNA from many different sources produced the same general X-ray diffraction pattern, the pattern's diffuse nature gave no solid clues as to whether the underlying RNA structure contained one or two chains. By early 1956 I decided to change my focus from X-ray studies on RNA to biochemical investigations on ribosomes when I began teaching in the fall at Harvard. Also then seeking a more tractable challenge was the Swiss-born biochemist Alfred Tissières, then studying oxidative metabolism at the Molteno Institute in Cambridge. He had already briefly dabbled with ribosomes from bacteria and liked the idea of our seeking out how they work across the Atlantic in the other Cambridge.

Alfred came from an old Valais family that long owned a bank in Sion. When he was less than a year old his banker father tragically died during the great influenza epidemic of 1918. Much later a minor inher-

Alfred Tissières braves an airy traverse of the
Gilgit River in northern Pakistan in 1954.

itance let Alfred possess the sleek Bentley that he parked across the
Cam on land adjacent to the school for the famed King's College boys'
choir. An even greater source of pride than his car was Albert's election
to the British Alpine Club in 1950. His formidable ascents of the south
face of the Taschhorn and the north ridge of the Dent Blanche led to
an invitation to join the Swiss 1951 Everest reconnaissance expedition.
Regretfully he had to decline, giving priority to his research efforts in
the Molteno Institute that led, in 1952, to a research fellowship at
King's. Climbing, however, always remained essential to his psyche. In
the summer of 1954 he joined in the Alpine Club's reconnaissance of
Pakistan's Rakaposhi, at almost eight thousand meters high one of the
Karakoram's most daunting peaks.

   After I left for Harvard, my successor as the Unit's geneticist was to
be the South African–born Sydney Brenner. We first met when he was
working for a Ph.D. at Oxford following medical training in Johannes-
burg. In the spring of 1953, Sydney was among those to have come to
Cambridge to have a peek at our big molecular model of the double
helix. He entered our lives more importantly, however, during the
summer of 1954, when Francis and I were at Woods Hole on Cape

Cod, talking genetic codes with the Russian-born, big bang theoretical physicist George Gamow. Then learning bacterial genetics at Cold Spring Harbor, Sydney came to Woods Hole for several days, greatly impressing Gamow and Francis by his quickness to catch on to their ideas and to propose experiments to test them.

Gamow, then a professor at George Washington University, was first drawn to the double helix through his reading in the summer of 1953 of our second *Nature* paper on the subject ("Genetical Implications of the Structure of DNA"). By early 1954, his seemingly wacky initial ideas had crystallized into a precise mechanics for the genetic code by which overlapping groups of three nucleotides coded for successive amino acids along polypeptide chains. On an early May 1954 visit to Berkeley, where George was on sabbatical, I proposed that we form a twenty-person code-seeking club, one member for every amino acid. George instantly reacted positively, much anticipating designing a tie and stationery for our RNA Tie Club.

Though there was never a convention of all its members, "notes" that circulated within the RNA Tie Club greatly advanced thought about genetic codes. The most famous of these notes, by Francis, in time would totally change the way we thought about protein synthesis. In January 1955 he wrote to the RNA Tie Club correctly suggesting that amino acids, prior to being incorporated in polypeptide chains, would attach to small RNA adaptors that in turn bind to template RNA molecules. For each amino acid, Francis postulated, there must exist a specific adaptor RNA (now called transfer RNA). In the absence then of any experimental evidence for small RNA, much less their chemical binding to amino acids, even Francis could not long remain buoyant about his adaptors. Six months were to pass before he was to regain a manic mood, but this time it was over a 3-D model for collagen that he and Alex Rich built over the summer of 1955.

After Alex returned in December to his job at the National Institutes of Health outside Washington, D.C., Francis and I focused for the winter of 1956 on the structures of small spherical RNA viruses, outlining how their cubic symmetry resulted from the regular aggregation of smaller asymmetrical protein building blocks. How their single long RNA chains were organized with their polyhelical protein shells remained to be seen. Our last time as a team of two was at a Johns Hopkins University–organized symposium in mid-June 1956, entitled "The Chemical Basis of Heredity." Upon arriving at the Hotel Baltimore, Francis jubilantly pointed out that we had been assigned adjacent rooms in the top-floor presidential suite.

After that occasion staying at the top was to be a challenge we would have to face separately.

Remembered Lessons

1. Choose an objective apparently ahead of its time

Mopping up the details after a major discovery has been made by others will not likely mark you out as an important scientist. Better to leapfrog ahead of your peers by pursuing an important objective that most others feel is not for the current moment. The 3-D structure of DNA in 1951 was such an objective, regarded by virtually all chemists as well as biologists as unripe. One well-known scientist then toiling in DNA chemistry predicted that a hundred years would pass before we knew what the gene looked like at the chemical level. Before setting out, you need to figure out a new path by which to climb—or, even better, a new intellectual catapult that can potentially hurl you over crevasses seemingly too broad to be leapt over by experimentation. The model-building approach to the DNA structure in 1951 had the potential to let us get where we needed to go at a time when the more orthodox approach, limited to analyzing X-ray diagrams, was far from straightforward. Given Pauling's recent success using molecular modeling to find the α-helix, using this approach on DNA was far from outlandish; actually, it was a no-brainer.

2. Work on problems only when you feel tangible success may come in several years

Many big goals are truly ahead of their time. I, for one, would like to know now where exactly my home telephone number is stored in my brain. But none of my colleagues who think about the brain yet know even how to approach this problem. We might do very well by asking how the cells in the much, much smaller fly brain are wired so as to recognize the odor of a specific alcohol—that would be getting us somewhere.

I feel comfortable taking on a problem only when I believe meaningful results can come over a three-to-five-year interval. Risking your career on problems when you have only a tiny chance to see a finish line is not advisable. But if you have reason to believe you have a 30 percent chance of solving over the next two or three years a problem that most others feel is not for this decade, that's a shot worth taking.

3. Never be the brightest person in a room

Getting out of intellectual ruts more often than not requires unexpected intellectual jousts. Nothing can replace the company of others who have the background to catch errors in your reasoning or provide facts that may either prove or disprove your argument of the moment. And the sharper those around you, the sharper you will become. It's contrary to human nature, and especially to human male nature, but being the top dog in the pack can work against greater accomplishments. Much better to be the least accomplished chemist in a super chemistry department than the superstar in a less lustrous department. By the early 1950s, Linus Pauling's scientific interactions with fellow scientists were effectively monologues instead of dialogues. He then wanted adoration, not criticism.

4. Stay in close contact with your intellectual competitors

In pursuing an important objective, you must expect serious competition. Those who want problems to themselves are destined for the backwaters of science. Though knowing you are in a race is nerve-racking, the presence of worthy competitors is an assurance that the prize ahead is worth winning. You should feel more than apprehensive, however, if the field is too large. This usually means you are in a race for something too obvious, not enough ahead of its time to deter the more conservative and less imaginative majority. The presence of more than three or four competitors should tell you that your chance of winning is not only low but virtually incalculable since you are unlikely to have a detailed knowledge of the strengths and weaknesses of most of your competition. The smaller the field, the better you can size it up, and the better the chance you will run an intelligent race.

Avoiding your competition because you are afraid that you will reveal too much is a dangerous course. Each of you may profit from the other's help, and an effective dead heat that allows you to publish simultaneously is obviously preferable to losing. And if it happens that someone else does win outright, better it be someone with whom you are on good terms than some unknown competitor whom you will find it hard not to at least initially detest.

5. Work with a teammate who is your intellectual equal

Two scientists acting together usually accomplish more than two loners each going their own way. The best scientific pairings are marriages of convenience in that they bring together the complementary talents of those involved. Given, for example, Francis's penchant for high-level crystallographic theory, there was no need for me also to master it. All I needed were its implications for interpreting DNA X-ray photographs. The possibility, of course, existed that Francis might err in some fashion I couldn't spot, but keeping good relations with others in the field outside our partnership meant that he would always have his ideas checked by others with even greater crystallographic talents. For my part, I brought to our two-man team a deep understanding of biology and a compulsive enthusiasm for solving what proved to be a fundamental problem of life.

An intelligent teammate can shorten your flirtation with a bad idea. For all too long I kept trying to build DNA models with the sugar phosphate backbone in the center, convinced that if I put the backbone on the outside, there would be no stereochemical restriction on how it could fold up into a regular helix. Francis's scorn for this assertion made me reverse course much sooner than I would have otherwise. Soon I too realized that my past argument had been lousy and, in fact, the stereochemistry of the sugar-phosphate groups would of course move them to outer positions of helices that use approximately ten nucleotides to make a complete turn.

In general, a scientific team of more than two is a crowded affair. Once you have three people working on a common objective, either one member effectively becomes the leader or the third eventually feels a less-than-equal partner and resents not being around when key decisions are made. Three-person operations also make it hard to assign credit. People naturally believe in the equal partnerships of successful duos—Rodgers and Hammerstein, Lewis and Clark. Most don't believe in the equal contributions of three-person crews.

6. Always have someone to save you

In trying to be ahead of your time, you are bound to annoy some people inclined to see you as too big for your britches. They will take delight if you stumble, believing your reversals of fortune are deserved. They may reveal themselves only in the moment of your discomfiture: often you find them controlling your immediate life by, say, determining whether you will get your fellowship or grant renewed. So it always pays to know someone of consequence—other than your parents—who is on your side. My hopes to go for broke with DNA by going to Cambridge would have gone nowhere if my phage-day patrons, Salvador Luria and Max Delbrück, had not come to my rescue when my request to move my fellowship from Copenhagen to Cambridge was turned down. I was then judged, not without cause, to be unprepared for X-ray crystallography and urged to move instead to Stockholm to learn cell biology. Immediately John Kendrew offered me a rent-free room in his home, while Luria, through a personal connection, got my fellowship extended for eight months. Soon after, Delbrück arranged a National Foundation for Poliomyelitis fellowship for the succeeding year. In finding the funds that kept me in Cambridge, Luria and Delbrück were hoping that my new career as a biological structural chemist would succeed and do them proud. But they fretted about my being too far from their fold, knowing that I would likely leave empty-handed from my long Cambridge stay. The second year of my fellowship was, in fact, to be spent at Caltech, giving me at least a measure of security in the event the DNA structure was solved by others. In leaving one field for another, it never makes sense to burn your past intellectual bridges at least until your new career has taken off.

7. MANNERS PRACTICED AS AN UNTENURED PROFESSOR

THE HARVARD to which I moved in the fall of 1956 thought of itself as the best university in the United States. Most certainly it was the oldest, and with its endowment the largest of any university's, it saw no reason not to have the most distinguished faculty of any institution on the planet. Before any tenure appointment, a group of eminent experts in the field were assembled to advise the president as to how the proposed candidate ranked among peers worldwide. The use of such ad hoc committees dated from the administration of James Conant, a distinguished organic chemist and only the second scientist ever to lead Harvard. Taking over from Lawrence Lowell in 1933, he presided for twenty years, resigning in 1953 to serve as U.S. high commissioner and later ambassador to Germany. Deeply involved in the military-related science that helped the United States win World War II, he seized upon the improvements in the nation's scientific capability to raise the bar correspondingly at Harvard's mathematics, physics, and chemistry departments.

The Harvard biology faculty contained several world-class scientists, in particular the vision biochemist George Wald and the evolution authority Ernst Mayr. But too many of its faculty had pedestrian outlooks incommensurate with the quality of most Harvard students. All too typical was the Biology Department's uninspired introductory course. It abounded in dull facts for its largely premedical enrollees to memorize. One year its abject dreariness provoked the student-written "Confidential Guide" to suggest that one of its instructors might do well to shoot himself.

Unlike Caltech, where genetics was the dominant biological discipline, Harvard's department, then chaired by the pedantic amber insect specialist Frank Carpenter, did not treat one field of biology as any more important than another. Together with his forlorn assistant, the former Rhodes scholar Orin Sandusky, Carpenter lumberingly oversaw the department's day-to-day activities in the massive five-story Biological Laboratories. It was built in the early 1930s in brick textile factory style, much of the money for its construction coming from the General Education Board of the Rockefeller Foundation, whose members wanted the benefaction to promote research as opposed to teaching. The nonexistence of a Biolabs lecture hall big enough for large biology classes was thus not a mistake but a matter of principle.

By the time the construction of the Biolabs started in 1932, the Depression had arrived and funds to outfit the north wing never materialized. Twenty-five years later, this wing's long empty factory-like floors suggested themselves to me as more than sufficient space for DNA-based biology to thrive at Harvard if the university was so inclined. Equally important to this objective, many senior faculty members were on the verge of retiring. Their large square corner offices, connecting to secretarial areas, themselves big enough for professors in less prestigious institutions, would soon be free. No lunch-room existed within the Biolabs either, and at noon the notables set off for the Georgian-style Faculty Club on Quincy Street. There they invariably lunched by themselves around the same rectangular table just inside the main dining room. Administrative minutiae, not ideas, dominated most conversations, with food chosen from a menu featuring horse steak, a proud holdover from wartime's austerity. Off the main dining room and usually entered by its own outside entrance was a separate room for women guests. Then there were effectively no women on Harvard's Faculty of Arts and Sciences.

With its corridor walls seemingly unpainted for at least a decade, the Biolabs' only sparkle came from the two enormous bronze rhinoceros that flanked the main entranceway. They had been sculpted by a talented friend of President Lowell's, who also designed the friezes of wild animals that ran above the courtyard. The vision of biology these figures conveyed meshed well with the mission of Harvard's nearby Department of Geographical Exploration, its building still topped by the radio antenna once used to keep in touch with members out beyond the fringes of Western civilization. But that department no longer existed. Rumor had it that President Lowell had been horrified to learn that several of its members were homosexuals. So its handsome one-story brick edifice was now the center of Harvard's Far Eastern studies, where the savvy John King Fairbank and Edwin O. Reischauer held sway.

Even closer to the Biolabs along Divinity Avenue was the Semitic Museum, donated by the banker Jacob Schiff at the end of World War I to encourage the study of ancient Jewish culture. But now most of its facilities were occupied by the Bob Bowie– and Henry Kissinger–led Harvard Center for International Affairs (HCIA), whose acronym encoded the identity of its secret government funder, which had an interest in training Harvard's students as the possible future leaders of the free world.

On the far side of the elm-lined grassy courtyard in front of the Biolabs stood what once had been the principal dormitory of the Harvard Divinity School. Ralph Waldo Emerson was said to have lived there early in the nineteenth century. But such historical facts mattered little to James Conant, under whose presidency the Divinity School's long minor role in Protestant theological training had withered almost to extinction. Just before my arrival, religion at Harvard was given a new lease on life through the appointment of Nathan Marsh Pusey as its next president. Born in Iowa in 1907, Pusey had studied classics as a Harvard undergraduate and had obtained his Ph.D. there at the age of thirty. After teaching at Lawrence, Scripps, and Wesleyan colleges, he returned to Wisconsin as president of Lawrence College in 1944. There he was to achieve postwar renown by speaking out against his state's junior senator, Joseph McCarthy. In choosing him as James Conant's successor, the five members of the Harvard Corporation saw themselves reaffirming the importance of a strong moral overtone in higher education. They were not unduly concerned that Pusey did not have the intellectual distinction to be a member of its faculty. Later they were to silently realize that his writings never sparkled and that his addresses to both students and faculty were occasions of neither enlightenment nor inspiration. And when they inevitably built a library in his memory, it was a below-ground structure intended to store archives.

To Pusey's credit, he accepted the Corporation's advice to appoint a first-class dean of the Faculty of Arts and Sciences. Whether he knew that in McGeorge Bundy he was choosing someone who would out-class him on virtually any occasion they were together, we will never know. A Boston blueblood by birth, Bundy came to Harvard via Groton and a brilliant undergraduate career at Yale. At Harvard he was initially one of the elite junior members of the Society of Fellows, later joining the Government Department and securing tenure by the time he became Harvard's most important dean. All appointments to the Faculty of Arts and Sciences would be administered by him, and it was he who would choose the ad hoc committees whose deliberations he and President Pusey invariably attended.

It is highly unlikely that Bundy had any role in Pusey's ill-fated decision, made in his second year as president, to deny the request of a Jewish student to be married in Harvard's imposing Memorial Church, built in the 1920s in memory of the American fallen of the First World War. In so doing, Pusey aroused the wrath of his faculty. A prominent delegation came to his office to tell him that Harvard's church should be open to those of all faiths, not restricted to Christians. It was a grievance rooted in history. Many years before, Jews had been effectively blackballed from faculty positions. Those faculty who had come to the president's office were determined that such bigotry as had stained Harvard's past would not corrupt its present. Sensing a fight that would effectively destroy the moral authority for which he was appointed, Pusey reversed his edict and the incident soon faded from view.

For Harvard's president, however, it was deeply wounding to be told that his initial response, which he regarded a reaffirmation of his institution's long Protestant heritage, was an expression of anti-Semitism.

From that moment on, Pusey never again saw his faculty as allies and became socially isolated from them during his remaining eighteen years as president. For friendship, he and his wife, Anne, would turn to the governing boards. They became summer residents of Seal Harbor on Mt. Desert Island, Maine, close to the home of David Rockefeller, soon to become chairman of Harvard's Board of Overseers. Both leaders felt similarly about the importance of religion, with Rockefeller making a major gift to strengthen the faculty of the Divinity School.

My decision to leave Caltech for Harvard was facilitated by a growing friendship with the chemist Paul Doty, whose laboratory in Gibbs Lab was just across Divinity Avenue from the Biolabs. Paul, trained initially as a physical chemist and then a polymer chemist, began physical-chemical studies of DNA only after moving to Harvard in 1948. Eight years older than I, he had just become a full professor when I arrived at Harvard. Fortunately for me, he was one of a handful of key faculty to whom McGeorge Bundy regularly turned for advice. So while many Harvard biologists remained uncertain as to whether I belonged in their department or in chemistry, Bundy, through Paul, knew I was a true biologist and hoped I'd help make the biology department into one comparable in stature to the ones in chemistry and physics.

Reassuring me that my academic life would not be totally at the whim of old-fashioned biologists was the recent formation of the Committee for Higher Degrees in Biochemistry, whose members were to be drawn from suitable individuals in the Biology and Chemistry departments. As a member from Biology, I would help choose the first class of graduate students and advise on appropriate courses for their first year. My first research student, Bob Risebrough, had been admitted as a Biology Department graduate student. As an undergraduate at Cornell, his main focus had been ornithology. Now he was excited by DNA, and his best introduction to it, I decided, might be to do a thesis on the properties of phage φx174, then reported to be much smaller than any other known phage. Its DNA molecules might be correspondingly smaller, thus perfectly suited to Paul Doty's physical chemistry instrumentation. Later I put my first biochemistry graduate student, Julian Fleischman, to work on the task of establishing the sizes of the DNA molecules in the much bigger T2 phage. Conceivably each T2 particle contained several DNA molecules held together end to end by protein linkers. Studying them might provide a good model for how DNA is arranged in the chromosomes of higher cells.

When Paul Doty ominously told me that promotions to tenure were often decided based on teaching evaluations, I realized I couldn't give the old-fashioned biologists a reason to suggest I might be better suited to a pure research institution or medical school. My attention focused sharply in my first months on my teaching assignments. Invariably worried that I would not have enough material memorized to occupy the next instructional hour, I meticulously outlined all my coming lectures. By doing so, I could offer my virus course students, largely advanced undergraduates, copies of the outlines, thereby relieving them of the need to take notes. Few students, however, availed themselves of this opportunity, continuing to be so sophomorically absorbed in note taking that their faces never revealed whether they were following my arguments. Fortunately, not too many stumbled in the hourlong midterm exam. And remembering the long-term benefits that had accrued to me at Indiana University from writing term papers on personally intriguing research topics, I asked them to write ten to fifteen pages on something in the course that particularly caught their fancy.

Initially I hoped to effect my social integration into the Harvard scene by living in one of the large undergraduate residence halls. Called houses, their creation realized President Lowell's wish to establish between Harvard Yard and the Charles River replicas of the Cambridge and Oxford colleges. As such, they would have young unmarried "tutors" living in specially designed suites. I asked my departmental chairman, Frank Carpenter, about the possibility, and he advised I try Leverett House, where the master was the embryologist Leigh Hoadley. Though he had long given up even a pretense of being a scientist, I saw no reason to assume Leigh would prove equally useless as a house master. All too soon, however, I discovered that the "bunny hutch," as Leverett House was then known, was never a first choice for undergraduates and that its so-called high table was the antithesis of what I had known in Cambridge. We ate the same uninspired food as the undergraduates, and conversation followed the lead of Master Hoadley, incapable of either levity or deep thought.

The ersatz high table might have mattered less if I had been provided with adequate living quarters. But my so-called suite did not look out on the Charles, its only view being to the opaque bathroom window of the master's apartment. My psyche was not helped by Hoadley's later admission that he might have given me accommodations more appropriate for a dog. I saw no reason to immediately let him know when I moved to a one-room flat carved out of a large house on nearby Francis Avenue. My first lab assistant, Celia Gilbert, daughter of the radical journalist I. F. Stone, had told me that her father's friend Helen Land had a vacancy nearby. It was one of several such small flats that I later realized were rented mainly to individuals with leftist connections. As I moved in, the journalist-to-be Jonathan Mirsky was moving out of the same building. His apartment was later occupied by a government graduate student, Jim Thomson, whom I would later meet when he became a member of the National Security Council.

In coming to Harvard still unmarried, I was more than conscious of goings-on at the once quite separate women's college, Radcliffe. Its residence halls were less than a mile away, and after the war classes at both colleges became entirely coeducational. Only the undergraduate Lamont Library remained out of bounds for women. How to go about meeting Radcliffe girls was not obvious, as their occasional mixers, then called jolly-ups, never seemed to bring forth the ones you would want to be seen with. Luckily, the geneticist Jack Schultz had a daughter, Jill, whom I had known earlier in Cold Spring Harbor, and who was now a Radcliffe senior living in a small wooden house off campus on Massachusetts Avenue. Soon I was to meet several of her housemates and gradually acquired the confidence to show up unannounced for after-dinner coffee.

Eating by myself in the faculty club was never an event to be anticipated and so I always greatly welcomed invitations to dine with the Dotys, now living less than a thousand feet from Paul's lab in a huge mansard-roofed house on Kirkland Place. Equally important in maintaining my morale were dinners at Wally and Celia Gilbert's equally proximate flat. We had met the year before at Cambridge University, where Wally had gone from Harvard as a young theoretical physicist. Knowing that they soon would be going back to Harvard upon completion of Wally's Cambridge Ph.D., I offered Celia, who had been an English major at Smith, a job at my lab starting in the fall. With Celia about, even routine lab manipulations became moments of conversational mischief. But after only four Biolab months, she was struck with mononucleosis. Her illness ended her tenure in my lab and, perhaps as a small consolation, the anxiety she suffered when called upon to dilute phage solutions by factors as big as a million.

Subtle conversational moments returned in March when Alfred Tissières, with his Bentley, arrived from Cambridge. Soon finding himself a room in a house off Brattle Street, he took on the task of finding a lab technician to replace Celia. Happily, Kathy Coit, whose parents were now housing Alfred, expressed interest in joining us. Finding her not only intelligent but also an enthusiastic rock climber, Alfred persuaded her to become our factotum. Though this was her first exposure to science, Kathy's cheerful common sense soon made her indispensable to our day-to-day lab progress.

Covering Alfred's salary was a grant that I had obtained from the National Science Foundation to study the ribosomes of bacteria. Those funds also allowed us to buy a preparative Spinco ultracentrifuge needed to spin them away from other bacterial components. A more expensive analytic Spinco that could measure how fast ribosomes sedimented was needed, too, but my grant wouldn't stretch that far. Luckily, we had one at our disposal thanks to the protein chemist John Edsall on the floor above.

Most evenings I would be back in the lab, having already spent the daylight hours there. After hours, we were required to sign the night watchman's sign-in book. There was no good reason for its existence except catching an errant husband in a lie concerning his whereabouts of an evening. One night I entered and was pleased to find it had gone missing, with no untoward consequences for the building's proper function. More frustrating was the bolting of the departmental library when the dour librarian went home. Though faculty members had keys, graduate students didn't and could not search out

Harvesting tobacco mosaic virus in 1958. From left to right: Julian Fleischman, Kathy Coit, John Mendelson, and Chuck Kurland

journal references in the evenings or on weekends. My continued pestering for the department to pay students to guard the entrance finally led to that reform for the common good.

Nathan Pusey regularly opened Harvard's stately President's House to his faculty and their spouses for Sunday afternoon tea and cakes. Paul Doty urged me to sample such an occasion, and I semiawkwardly presented myself at the front door when my fall term lectures were nearing their end. Led by a maid into the main drawing room, where I introduced myself to the Puseys, I soon was passed along to talk with the late-thirtyish Swedish theologian Krister Stendahl and his equally youthful wife, Brita. A prize catch in Pusey's efforts to resurrect the Divinity School, Stendahl had a strong, angular, slightly distorted face that struck me as that of a troubled minister in an Ingmar Bergman film. Liking his reasoned openness to the complexities of human life, I nevertheless could not even affect interest in the Evangelist Matthew, about whom he had just completed a scholarly tome. Later, when

Anne Pusey moved to be near me, I felt much more at ease talking about my Chicago education and how fortunate I felt to be part of the Harvard scene. Simultaneously I tried to overhear what our president was saying to others. Later, as I walked out onto Quincy Street, I wondered whether any conversational gambit could possibly elicit from him an animated response.

Later during the monthly meetings of the Faculty of Arts and Sciences, I was no more successful at discerning the feelings that occupied what he considered to be his soul. We always turned to Bundy for hints of what was coming next. Pusey seemed to come to life only when presenting honorary M.A.'s to those newly tenured faculty whose actual degrees had been conferred elsewhere. Through this gesture of sanctification, Harvard saw itself as ensuring that all faculty felt equally valued.

My social life at Harvard still left much to be desired. I had flown to London just before Christmas, and then for the New Year had gone by train up to the home of Dick and Naomi Mitchison on the Mull of Kintyre in the Scottish Highlands. My first visit to their Carradale House had been five years before, when I was invited by their youngest son, Avrion, then doing his Ph.D. thesis in Oxford on the immunological response. Av's mother was a distinguished writer of leftist persuasion, so I could again count on being part of an intellectual house party that featured long walks over boggy moors, heated conversations much more about politics than science, and hearty but never inspiring food. Still, I knew it would be much more fun than going back to the small home in the Indiana sand dunes to which my parents had moved after my sister's graduation from the University of Chicago. I would later regret not having been a more dutiful son when my mother, only fifty-seven, died of a sudden heart attack not long after the holidays. She never had the pleasure of visiting Harvard to see me as a member of its faculty. When I went home for her funeral, I could see my father was unlikely ever to recover entirely from her unexpected death.

At the end of July, I was happy to be able to bring my father along to the Isle of Skye, where I was to be the best man at Av Mitchison's wedding to Lorna Martin. It was my first chance to meet Av's intellectually powerful research supervisor, Peter Medawar. He came up from London with his strong-willed wife, Jean, and fetching bright daughter, Caroline, then intending to escape from much unwanted parental dating advice by going up to Cambridge. In the middle of the reception, I had no difficulty in spiriting Caroline away for a long car ride through the wild beauty of Skye, remaining absent long enough for Peter and Jean to grow worried that Caroline and I might have found each other perfect. But she had other plans for the next few weeks. And after putting my father on a plane back to Chicago, I anticipated intersecting in Tuscany with a Radcliffe girl I knew from the off-campus house on Massachusetts Avenue, who was traveling in Europe. Several letters from the prior locales on her itinerary gave me to believe that she would greet me warmly when our paths finally crossed in Assisi. But as we looked up at the Giotto frescos on the walls of its fifteenth-century basilica, I sensed that her affections were already subscribed; I later learned she was smitten with a young classics instructor.

Even before I went off to Europe, momentum was building for the appointment of a super geneticist to the biology faculty. Bagging such a star had been on Harvard's agenda since the late 1940s, when an offer failed to lure Tracy Sonneborn away from Indiana. Now the introductory course was being taught by Paul Levine, whose recent promotion to a tenured position had been a dicey matter. In so proposing him, the senior biology faculty had to emphasize his highly praised teaching, since his middling research on *Drosophila* was far from noteworthy. Given Bundy's known determination to prevent his out-of-date biology faculty from perpetuating their inherent mediocrity, the promotion had hints of a compromise between him and Pusey. The letter from University Hall to the Biology Department stated that Levine's appointment was conditional upon the department's next tenured slot being reserved for a geneticist whose research was world-class. Initially I feared that they might somehow find a candidate who, though clever, did not yet feel it necessary to think in terms of the double helix. Fortunately, I could not have been more wrong.

The subsequent departmental committee made the Purdue University phage geneticist Seymour Benzer, already my close friend, number one on their list. Decisive was the maize geneticist Paul Mangelsdorf's highly positive reaction to Benzer's recent talk before the International

Congress of Genetics in Toronto. Within days of Seymour's coming to Harvard to talk about his fine-structure genetic map of the r2 gene of phage T4, my senior biology colleagues voted unanimously for his appointment. Seymour, of course, had been earlier apprised of the department's intentions. Paul Doty and I both told him that it was virtually impossible for any properly constituted ad hoc committee not to back his addition to the Harvard faculty.

Alfred Tissières was now Bentley-less. In July he was to marry in Colorado the equally strong-willed Virginia Wachob, a girl of Scottish descent from Denver, whom he first met at Caltech during a year away from King's. Supporting both a wife and a Bentley, which soon might need a major engine overhaul, would not be possible given that Alfred's Harvard salary was slightly lower than mine. The Bentley now belonged to a much more handsomely rewarded law school professor.

With the assembling of the ad hoc committee taking longer than expected, it was early February 1958 before Seymour had his formal offer and an invitation back to Harvard to meet with Bundy. Then he wanted reassurance that DNA-based biological thinking would have the continued strong support of the administration. To my distress, Seymour didn't immediately accept, implying worry at having greater teaching responsibilities than at Purdue. And so it was a great relief when Bundy happily called me early the next week with news that Seymour's letter of acceptance was on his desk.

Going off soon thereafter to the University of Illinois, as its George D. Miller Visiting Lecturer in Bacteriology, I could finally rest assured that the days of backward thinking in Harvard's Biology Department were numbered. My visit was arranged by Salva Luria, who by then had been a professor at Urbana for more than five years. My lectures on macromolecular replication and cell growth were a preview of ones I wanted to deliver later to Harvard undergraduates. Over the three weeks of my visit, I enjoyed much stimulation from the Urbana science scene. Especially enjoyable was talking with the diminutive, manic Sol Spiegelman, then also focusing much of his research on ribosomes.

Flying back to Boston on an intellectual high, I came back down to earth at Harvard with a thud. Seymour Benzer now claimed a heart

THE GEORGE A. MILLER VISITING LECTURER IN BACTERIOLOGY

Dr. James D. Watson
HARVARD UNIVERSITY

SIX LECTURES ON THE GENERAL TOPIC
Macromolecular Replication and Cell Growth

1. **NUMBERS AND FIGURES**
Thursday, February 13

2. **SOME PRINCIPLES OF MACROMOLECULAR STRUCTURE**
Tuesday, February 18

3. **CELLULAR REPLICATION**
Thursday, February 20

4. **GENETIC SIGNIFICANCE OF THE NUCLEIC ACIDS**
Monday, February 24

5. **NUCLEIC ACID REPLICATION**
Wednesday, March 5

6. **THE CONTROL OF PROTEIN SYNTHESIS**
Friday, March 7

ALL LECTURES
4:00 P.M., 116 EAST CHEMISTRY BUILDING

condition that forced him to reverse his decision to come. Staying in Purdue, with its almost nonexistent teaching responsibilities, would be much less taxing, and he had to think of his health. The disappointment might have been unbearable if Av Mitchison had not happened to be in residence for the spring term, giving an advanced course on immunology. To my delight, he and his new wife, Lorna, together with Alfred Tissières and I, were temporarily occupying the Dotys' big house on Kirkland Place while the Dotys spent Paul's long-overdue sabbatical in the other Cambridge. Parked next to my MG TF beneath the Dotys' main bedroom was Alfred's consolatory sleek new Alfa Romeo sedan.

Had it not been for Doty's occasional flights back to oversee his ever-growing lab group, my direct line to McGeorge Bundy would have been cut off when I most needed it. With virtually no warning, I learned that an ad hoc committee soon would assemble to consider simultaneously the promotions from assistant professor to tenured associate professorship of Edward O. Wilson and myself. That I was to be considered a year prematurely was not at all the original intent of the Biology Department. Their concern primarily was Wilson, whom they needed to promote to keep him from accepting the same terms from Stanford. After undergraduate education as a naturalist in his home state of Alabama, Ed had moved north to Harvard to pursue a Ph.D. Proving his brilliance during his initial studies on ants and their behavior, he became a junior fellow, then considered the best stepping-stone to an eventual permanent position on the faculty. Since my arrival, we seldom had reason to speak: I was a midwesterner, he a southern boy; he was par excellence a naturalist, while I knew nothing about ants, having by then lost all my earlier interest in animal behavior. But the vast museums of Harvard's past glory were not to vanish, and it appeared that Wilson might very well have the intelligence and drive needed to move Harvard's evolutionary tradition into the future.

Since my research achievements were already internationally noted and no one could say that I had either shirked or botched my teaching responsibilities, Paul Doty felt that fairness dictated that the Biology Department now also make up its mind as to whether it wanted me as a permanent member. Bundy happily agreed and unilaterally informed the Biology Department that he and Mr. Pusey wished to consider my appointment as well as that of Ed Wilson. With Wilson's offer from Stanford needing an answer soon, an ad hoc committee was formed even before a department vote, just before my thirtieth birthday on April 6. Through one of its members, the highly perceptive Rockefeller Foundation science executive Warren Weaver, Paul quickly learned that the verdict was thumbs up for both of us. By then Bundy had already told the Biology Department that he had President Pusey's permission to promote me as well as Wilson. So Frank Carpenter assembled his senior biology professors the next day to see whether they would concur with the ad hoc committee's decision.

I was more than worried that one too many of those dinosaurs would vote against me. In fact, a majority of them did, opting to postpone for one year the decision on my promotion. This I heard from Ernst Mayr, who wisely didn't identify those against me, and I couldn't contain my outrage. Retreating to the Doty house before I used the

F-word in front of too many biology graduate students, I agitatedly awaited dinner with Paul and the Berkeley zoologist Dan Mazia. Over dinner Dan tried to console me by saying that at Berkeley they never would have tried to stave off what was so obviously inevitable. Paul Doty, trying to rescue his dinner party from the pall hanging over it, reassured me that the game was not over until it was over. McGeorge Bundy's power would be badly eroded if he let one of his second-rate departments defy him. Paul counseled me to try to refrain, at least for the time being, from further vulgar diatribes against my biology colleagues.

The subsequent weekend was inevitably tense, as I waited to see the color of the smoke emerging from University Hall. To my great relief, Bundy did play hardball, telling Frank Carpenter that no more tenured appointments or discretionary funds at the dean's disposal would go biology's way until they promoted me. Quickly those professors who only several days earlier were strongly opposed to me now implied they had acted too hastily. Upon further reflection they could now enthusiastically accept the ad hoc committee's recommendation.

Relieved not to have to consider offering myself to the Chemistry Department, I couldn't find it in me to gloat. But it was hard not to appreciate Seymour Benzer's later comment that the imbroglio attending my promotion made him even more certain that he had done the right thing turning down Harvard. Life was too short to share a department with so many prima donnas whose meager accomplishments scarcely justified even the status of has-been. Still, I did not regret moving to Harvard. More and more I was learning that the quality of your students matters much more than that of your faculty colleagues. In that regard Harvard couldn't be faulted.

Remembered Lessons

1. Bring your research into your lectures

In the fall of 1956, there was simply not enough known about DNA to organize a whole course around it. So I opted to talk about DNA in the context of a course on viruses, wherein I could compare the elegant experiments of the phage group with the old-fashioned approaches of plant and animal virologists. Graduate students self-selected according to their attraction to my molecular messages. Reading through their term papers, I could also spot those who zoomed in on important issues and did not waste pages of type on observations of no consequence.

2. Challenge your students' abilities to move beyond facts

Asking bright students to merely regurgitate the facts or ideas of others does not prepare them for the world outside classrooms. So my exams increasingly featured questions that assessed the plausibility of hypothetical headlines from the *New York Times* or *Nature*. For example, should they believe claims that a virus had been found that multiplied outside cells in a medium solely composed of the small-molecule precursors to DNA, RNA, and protein? Any student answering yes would have missed the essence of my course and so been advised not to choose a scientific career. Happily, no students failed to answer correctly.

3. Have your students master subjects outside your expertise

The best way to prepare your students for the independence they all want is by seeing that they are exposed to peripheral disciplines and to the technologies needed to move from the present to the future. During the late 1950s when we aimed to discover how information encoded within DNA molecules is expressed in cells, the answers had to be sought at the molecular level. So it was a no-brainer that I should have my prospective students acquire strong backgrounds in chemistry to complement my strengths as a biologist. During their first graduate year I made sure that they took rigorous courses in physical and organic chemistry. They might later use only a small fraction of this expertise, but they would never feel unqualified for experimentation at the molecular level.

4. Never let your students see themselves as research assistants

It makes sense to have your students pursuing thesis objectives that genuinely interest you. At the same time you should take care that they never see themselves as working primarily for your professional advancement. Students function best when they can be assured of enjoying most of the credit for their efforts. After they came into my lab, generally only a month or two would pass before I backed away from their daily progress. I then let them work at their own pace and come into my office when they had results, either positive or negative, that I should be aware of. You know that they have become truly independent when they give thoughtful seminars before their lab peers. Novice speakers can profit from taking their licks when their conclusions go beyond what is justified by their data. Nothing banishes illogical conclusions from one's brain like the need to present them to others. Later I made it a point that my name never be included with theirs on research papers emerging from their experiments.

5. Hire spunky lab helpers

As an untenured scientist, most of your nonsleeping hours are spent in lab-related activities. Those working with me were effectively my surrogate family, with whom I would eat many meals and go to the beach or go skiing. So when hiring assistants to help with more routine lab management, I wanted to surround myself with faces that laughed at the right times and whose inherent positive outlook would be a calming influence when our experiments went nowhere. The best to have around were unmarried people of my own age; not yet saddled with family responsibilities, they were therefore not obliged to keep strictly regular hours. They could be called on for help in the evening hours or on weekends when we wanted our answers fast. In return, I treated them more as friends than as employees and didn't expect them to hang around when there was nothing particular to do.

6. Academic institutions do not easily change themselves

Most academic battles involve space or faculty appointments and promotions. All too often, academic life is a zero-sum game, with an equivalent loser for every winner. Sadly, most academic department heads and deans do not display long-term consistency, often maintaining their own academic power by giving to a professor what he or she was denied the year before. Before I went to Harvard, Leo Szilard told me that it moved only lethargically, an assessment based no doubt on his never having been asked to join its Physics Department. But he was also familiar with academia's general love of orthodoxy and warned that I should be realistic about how much change I could expect to see in a place as fossilized as Harvard's Biology Department. His pessimism would have been dead on had it not been for McGeorge Bundy's determination to see through a radical upgrade of biology at Harvard. University leaders with such strong convictions are rare.

8. MANNERS DEPLOYED FOR ACADEMIC ZING

UPON the return of the Dotys from England, I needed a new place to live and luckily stumbled upon a vacant one-bedroom flat on the thousand-foot-long Appian Way, less than a five-minute walk from Harvard Square. Appian Way runs between Garden and Brattle Streets and is bordered on its northern side by Radcliffe Yard, where once virtually all of Radcliffe's classes were taught by Harvard professors. But with the disappearance of separate classes for women, the former classroom space in the Yard's Longfellow Hall was now used by the School of Education. Soon after my moving onto Appian Way, the Education School was to expand across it, in the process tearing down all its modest wooden homes except for number 10—the mid-nineteenth-century house, long owned by the Noon family, in which I occupied second-floor quarters. I regularly wrote out my bimonthly checks to Theodore W. Noon, who had been at Harvard at the turn of the century and an instructor at Lawrenceville before the first Great War. Long retired from teaching, he was nearing eighty, and would eventually live to almost a hundred.

I was pleased to be told that among former 10 Appian Way tenants were the writers Owen Wister and Sean O'Faolain. The building's ancient central heating system was less charming. In winter I routinely needed an electric blanket to sleep. My flat's Spartan features were made for inexpensive furniture I bought from the Door Store on Massachusetts Avenue on the way to Central Square. Soon to complement its simplicity was a big-planked New Hampshire harvest table that I found in an antiques store near Falmouth on Cape Cod. I hoped its inherent elegance would inspire some Radcliffe girls to test their cooking talents in my tiny kitchen.

By early fall I had lost all hope that the astute geneticist Guido Pontecorvo would move from Glasgow to fill the senior geneticist's slot turned down by Seymour Benzer. Six months before, the ad hoc committee called to look over Benzer had also judged Ponte's accomplishments worthy of a major offer, a conclusion simultaneously reached by the electors of the soon-to-open chair of genetics at Cambridge. But Ponte had strong attachments to Glasgow, where his colleagues had solidly stood by him during the difficulties that followed his physicist brother Bruno's sudden flight to Moscow just before he was to be charged with treason for passing atomic bomb secrets to Soviet agents. Ponte, in turn, stood by his friends by refusing to go to either Cambridge.

That fall, knowing that I would be teaching undergraduates in the spring, I gave a graduate-level course on the biochemistry of cancer. During my previous winter weeks at the University of Illinois, I learned that uncontrolled cell growth should be the province of the biologist as well as the medic. There the biochemist Van Potter, from the University of Wisconsin, gave an evening colloquium on cancer, making me aware that at any given time most adult animal cells are not undergoing cell division. To duplicate their chromosomes and divide, these cells must receive molecular signals that initiate chains of events culminating in the production of enzymes involved in DNA synthesis. In contrast, cancer cells very likely do not require external signals to enter into cycles of growth and replication. Future searches for such intrinsic mitosis-inducing signals would quite likely involve the already long-known tumor viruses. Upon infecting so-called nonpermissive normal cells, they do not initiate rounds of viral multiplication but rather transform the healthy cells into their cancerous equivalents. The Shope papilloma virus, already known for more than two decades to induce warts on rabbit skin, particularly intrigued me. Only a few genes were likely to be found along its tiny DNA molecules, made up of a mere five thousand or so base pairs.

Over the past summer at the Marine Biological Laboratory in

Woods Hole, I had met the biochemist Seymour Cohen, then very upbeat about his lab's recent discovery that T2 phage DNA contains genes that code for enzymes involved in DNA synthesis. Initially I did not attach the proper importance to Cohen's finding, but I abruptly changed my mind several months later when preparing a lecture on Shope papilloma tumor virus. Then I found myself asking whether its chromosomes, like those of T2, also code for one or more proteins that trigger DNA synthesis. Since at any given moment the vast majority of adult cells do not contain the large complement of enzymes necessary to make DNA, conceivably all DNA animal viruses had genes whose function was to activate synthesis of these enzymes. Even more important, these genes, if somehow integrated into cellular chromosomes, might make the respective cells cancerous. I was in a virtually manic state as I presented these thoughts to my class. At last I understood the essence of a tumor virus. Over the following days, however, I realized that my excitement did not infect those about me. Only those who come up with the seductively simple ideas initially get hysterical about them. Everyone else demands experimental proof before joining the conga line.

I was still high on my theory by the time of a Sunday cocktail party given early in 1959 by David Samuels, the British-Israeli chemist who had recently arrived at Harvard as a senior postdoc under the bio-organic chemist Frank Westheimer. David was in line to be a British lord, like his father and his grandfather, the first Viscount Samuels, an influential early backer of a Jewish state in the holy land. Now he was still saddened over the death of his cousin Rosalind Franklin from ovarian cancer the past April. They had seen each other often during their school years. But after he chose Oxford and she Cambridge for their education as chemists, their paths less often crossed. It was only now that I realized that Rosalind had come from no modest background. Had she wished, she easily could have moved into the world of moneyed society that David now so obviously enjoyed when not applying himself as a serious chemist.

The most striking of David's guests that afternoon was the Radcliffe senior Diana de Vegh. Quickly making a move to her side, I learned that her investment banker father, Imre de Vegh, was a Hungarian, while her mother was a Social Register Jay. As such, she had sampled both Brearley and Miss Porter's School before being admitted to Radcliffe, where she had effectively managed to avoid any science. Now she lived in one of the off-campus houses favored by her fellow private school friends. Her comings and goings were much less conspicuous than would have been those of an inhabitant of one of the large red brick dorms surrounding the Radcliffe Quad up Garden Street. David, in observing Diana give me her telephone number as she went off to another party, took immediate pleasure in telling me that she had earlier attracted the attention of Senator John Kennedy. His official car recently had been sent from Boston to fetch her upon one of his recent returns to check in on his Massachusetts constituents.

Soon I was to phone Diana to ask her to lunch after one of my Biology 2 lectures. This new course, a one-term offering, was intended for students already possessing some background in biology. They would have benefited most from a coherent series of lectures given by one person, in the manner of the long-successful Chemistry 2 taught by Leonard Nash. But my department had opted for Biology 2 to have four instructors, thereby ensuring a virtual potpourri of facts and ideas for its students to master. But with me as one of its four instructors, DNA was bound to be much talked about.

The theme running through all my talks was the need to understand biological phenomena as expressions of the information carried within DNA molecules. Many, and I hoped most, students must have been desperate after nine lectures presented by the physiologist Edward Castle. The tall, thin Castle was bright but sad, habitually seen hurrying home by bike early each afternoon to a wife long stricken with multiple sclerosis. His lectures were a time warp back into the thirties. After listening to his opening talk, I could not have preserved my sanity listening to another. Three weeks later, before the same one hundred students in the Geological Museum auditorium, my first words were a promise that they would hear nothing more about the rabbit. The loud laughs that roared back assured me that I had broken the ice. After my last lecture, which was to be about cancer, I took Diana de Vegh to Henri IV, the French restaurant on Winthrop Street just off Harvard Square, run by the formidable Genevieve McMillan.

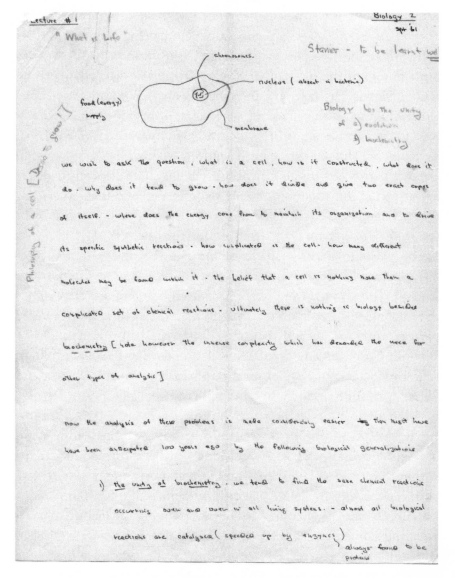

Course notes for Biology 2, Lecture 1: "What Is Life"

Genevieve had masterminded the transformation of a modest wooden house into a popular space to have stylish French food with conversationally inclined companions. Looking into Diana's big eyes, I was in high spirits, for my lectures had gone well, with my impromptu attempts at humor appreciated.

That term Francis Crick was at Harvard as a visiting professor of chemistry. Even now, six years after the double helix had been found, Cambridge University had yet to provide him and his new South African–born collaborator, the prodigiously clever experimentalist Sydney Brenner, decent research space. Their experiments were being done in a wartime hut erected next to the Austin wing of Cavendish Lab, where the DNA base pairs had come together. The Harvard Chemistry Department, always wanting the best, had just offered Francis a professorship, but without much expectation of his acceptance. Here they were right, as Francis soon got news that the Medical Research Council would provide funds for the construction of a new laboratory building expressly for molecular biology. In fine intellectual fettle because his long-ignored adaptor hypothesis was now a widely accepted fact, Francis conversed endlessly on the details of how specific amino acids first become attached to their respective tRNA molecules. When he and I were awarded the Warren Triennial Prize at Massachusetts General Hospital, my talk surely was much less convincing than his, as I took the opportunity to propose my as yet unproven theory that the essence of DNA tumor viruses was their possession of genes that initiated DNA synthesis.

Ribosomal particles were proving much more structurally complicated than we anticipated, and Alfred Tissières persuaded the Welsh protein chemist Ieuan Harris to come over temporarily from Cambridge to help us. We initially had thought the particles would have the molecular simplicity of small plant viruses. But from his first amino end group analysis, Ieuan saw that ribosomes contained many more proteins than he could effectively cope with using current separation methodologies. So he took solace for the remainder of the spring in sampling American beers still new to him.

My lab group's size was steadily expanding despite the unwanted fleeing of my first graduate student, Bob Risebrough, to sea on the Woods Hole oceanographic sailing boat *Atlantis*. Much more content at their first contacts with molecular biology were two new graduate students, David Schlessinger and Charles Kurland. After working with Alfred to more firmly establish the two-subunit composition of ribosomes, David briefly went to Caltech to see if Matt Meselson's CsC1 banding technique would reveal how long these subunits stayed together during multiple rounds of protein synthesis. That he came back empty-handed reflected his finding that even the ribosomal subunits are unstable in high levels of CsC1. But the visit was far from a total loss: at Caltech David met the girl that he would later marry. He also discovered David Zipser, a very disenchanted first-year graduate student eager for a fresh chance at happiness at Harvard.

Chuck Kurland's first results on ribosomal structure were not at all what I expected. He found single RNA chains in each ribosomal subunit, with those present in the larger subunit twice the length of those in the smaller subunits. Before his observations we had anticipated a variety of RNA chain lengths, reflecting their respective functions to convey information for different-sized polypeptide chains. A year later, Matt Meselson and Rick Davern discovered that once made, these ribosomal RNA chains were very stable under optimal conditions for protein synthesis. Yet at the time of our discoveries, François Jacob, Jacques Monod, and Arthur Pardee at the Institut Pasteur in Paris had evidence that the RNA templates for so-called induced enzymes had fleeting lives of only a few minutes. This apparent contradiction came to a head at a Copenhagen meeting in the late summer of 1959, at which Jacques Monod questioned the view that RNA had to be the template for all protein synthesis. Could, in fact, DNA molecules be the templates for the synthesis of the so-called induced enzymes? I rejected this hypothesis during my terse report of our ribosome discoveries. Unfortunately, my talk's only lively moment came at its start, and not from its content. Everyone in the front row brought out a copy of the *New York Times* to read—a brainchild of Sydney Brenner, who had long noticed with envy my ability to follow talks while simultaneously keeping abreast of daily events.

Earlier in the summer the Japanese biochemist Masayasu Nomura, then a postdoc in Sol Spiegelman's Urbana lab, came briefly to my lab on his way to the 1959 phage course at Cold Spring Harbor. He had spent the previous summer with me and Alfred characterizing abnormal ribosomes made under conditions of chloromycetin inhibition of protein synthesis. Now a year later and still unable to judge their biological significance, I suggested to Masayasu that he use his forthcoming phage course experience to look at the molecular form of the unstable RNA made during T2 infection. I had been long attracted to T2 RNA because its base composition was almost identical to that of the T2 phage DNA and possibly represented RNA copies of the information present in T2 DNA genes. Though potentially very important, the phenomenon eluded further characterization. What T2 RNA was should at last be revealed by the new techniques of sucrose gradient centrifugation, which required only small amounts of RNA and would potentially provide information about its function through measurements of the sedimentation rate of its radioactively labeled molecules.

We were greatly benefiting from David Zipser's transfer from Caltech. From the moment he arrived, David, soon to be called Zip, homed in on the sucrose gradient centrifugation procedure that had been recently developed at the Carnegie Institution of Washington biophysics lab. He was an indisputable asset, but a period of some acquaintance revealed that Zip's unhappiness at Caltech may have had a basis in his character: he did not play well with others. After a particularly inappropriate remark about another student's girlfriend, Zip was put on social probation: I told him to stay away from our afternoon tea and cookie sessions, a ban maintained for much of the year. Tellingly, no one came to his defense. His cause was not later helped by his cheering when Gary Powers's U2 was shot down in May 1960 and Eisenhower had to cancel his visit to Moscow later that month. Zip was the product of a New York communist family, fully aware that an easing of tensions between the United States and the USSR would give the family less purpose.

Everyone in the Biolabs was also now benefiting from the insect physiologist Carroll Williams's becoming the new chairman of biology. In July 1959, he got money from University Hall to repaint the corridor of the Biolabs, using bright shades of red, yellow, and blue at focal points. Much more important, McGeorge Bundy told Carroll that he wanted the Biology Department to propose five names for consideration by a super ad hoc committee to be formed in midwinter. Our department's failure over the past year to attract a distinguished geneticist had made Bundy realize that he would have to take the bull by the horns if he indeed wanted to force our Biology Department into the modern age. Not all the appointments were to be molecular biologists, but each should represent a significant advance over the past.

More a conciliator than I might have hoped, Carroll wanted the five candidates to have the department's unanimous backing. My elation at Bundy's proposal was tempered when I saw the department was soon to make offers to the animal and plant embryologists Aaron Moscona and John Torrey. As I believed embryology would remain an antiquarian pursuit until revived by better understanding of how DNA expression is controlled, I feared these offers would amount to two more nonwinners. More hope lay in the department's wish to hire an electron microscopist who could teach a modern cell biology course, and I was happy that they decided to try to pinch the spirited Keith Porter from Rockefeller University. For someone to teach modern microbiology, there seemed no one better than Mel Cohn at Stanford, a former protégé of Jacques Monod in Paris. Best of all, there was universal enthusiasm for trying to get Matt Meselson to defect from Caltech. Earlier Matt had mesmerized the department with a lecture on how his experiments with Frank Stahl demonstrated the semiconservative replication of DNA.

Conceivably, a major impediment to our effort lay in Jonas Salk's impending creation of a new high-powered research institute for biology at La Jolla. He bragged about its future oceanfront location on a stopover at Harvard in the fall of 1959 on his way to visit his son at school at Exeter. There he envisioned recruiting leading biologists, including me, to engage in cutting-edge research with no teaching responsibilities. Behind Jonas's utopian vision was Leo Szilard's desire to create a first-class environment that he could use as a permanent base for his simultaneous forays into biology and nuclear weapons politics. Leo strongly wanted Jonas as its head, believing his fame would make Hollywood Jewish money easy to get. My counterargument that Jonas's intellectual distinction was not much greater than Nathan Pusey's did not faze him. Leo countered that the bylaws of the new institute could be set up to put control of appointments and finances in the hands of its leading scientists. I was, in fact, more intrigued by the recent offer to be one of the founding biologists at a new University of California campus also to be located in idyllic La Jolla. It was to be adjacent to the land that Jonas coveted for his purposes.

My formal visit to see the proposed site of the U of C's new La Jolla campus occurred in early March 1960, timed to coincide with a gathering of prospective appointees to what later would be called the Salk Institute. Awaiting me at the San Diego airport was Jonas with a white stretch limousine to take us to a motel on the Pacific Ocean. The wheels, Jonas explained, had been provided by his longtime patron, the National Foundation for Infantile Paralysis, still then headed by former Franklin Roosevelt lawyer and confidant Basil O'Connor, who believed that only by such grand conveyance would Jonas get the respect due him for his role in stopping polio. Upon arriving at the oceanview site of the future Salk Institute, I saw that Matt Meselson and Mel Cohn also had been called to La Jolla by Jonas. I quietly told everyone that Meselson and Cohn would both soon also be receiving offers from Harvard. Unusually tense was a final luncheon jointly hosted by Jonas and the oceanographer Roger Revelle, head of the venerable Scripps Institution of Oceanography and the main instigator of the new U of C campus. Roger, who had married into the Scripps newspaper fortune, did not take kindly to Jonas's opening remarks that he and Jonas respectively play mama and papa at the occasion.

I left San Diego not at all tempted to move to its perfect climate. Instead I was keen to return to Harvard in anticipation of the impending visit of Peter, Jean, and Caroline Medawar. Three years had passed since we were all together on Skye and in the meanwhile Caroline had almost finished her Cambridge education. Peter was to give that year's Prather Lectures in the Biology Department. After the last of his three lectures was delivered, I made the mistake of inviting Caroline to go skiing in Vermont. Her awkwardness as a beginner was compounded by the seemingly fearless downhill glide of a young Radcliffe student I had also asked along. In a sulk at being outclassed by a girl who also had brains, Caroline went off to join her parents in New York, leaving me to realize once again that you shouldn't pay simultaneous attention to two girls.

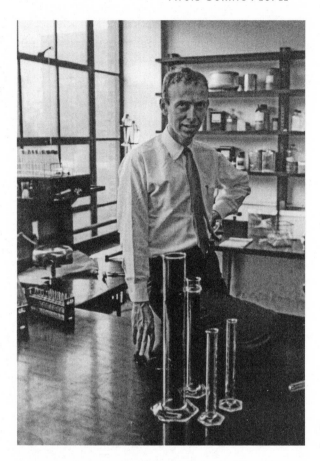

*In the Harvard
Biolabs*

Fortunately, a major conceptual breakthrough was soon to emerge from my lab. Late in 1959, as an editor of the newly founded *Journal of Molecular Biology*, I received a manuscript from Urbana by Masayasu Nomura, Ben Hall, and Sol Spiegelman on T2 RNA. My initial reading immediately led me to doubt its central conclusion, that the T2 RNA sedimented as if it were a special form of the small ribosomal subunit. Believing the experimental facts should be known widely, even if they were possibly wrongly interpreted, I accepted the manuscript for publication. Soon afterward I suggested to Bob Risebrough, just back from more than a year on the Indian Ocean, that he repeat the experiments done at Urbana in the hope that we might at long last see the RNA templates for protein synthesis. It had long mystified me why Bob had so precipitously vamoosed to sea. Now I took comfort in learning that it was not from my lab and its experiments that he had fled but from an entanglement with a married woman. Hoping this mess was now behind him, Bob wanted to get back in the game.

Within six weeks, Bob cut to the heart of the T2 RNA's special nature. Using sucrose gradients containing high (10^{-2}M) levels of Mg^{++} ions, all the T2 RNA was seen to have bound to the 70S ribosome complex of the big and small ribosomal subunits. In contrast, when Bob followed sucrose gradients containing lower (10^{-4}M) Mg^{++} levels, the T2 RNA sedimented as free RNA and not as part of either the smaller or larger ribosomal subunits. Excitedly we realized that ribosomal RNA never orders amino acids during protein synthesis. Instead their respective ribosomes are nonspecific "factories" in which the T2 RNA templates order amino acids during protein synthesis. That such messenger RNA had not been seen before reflected the fact that, in most cells, much more ribosomal RNA is made than messenger RNA. But following infection of bacteria by T2-like phages, all host-specific RNA synthesis stops. All the RNA molecules synthesized during phage infection are made on T2 DNA templates.

A week later I flew down to New York City, trying without success to resume my friendship with Caroline Medawar, who was spending the weekend with her parents at Rockefeller University. Over the same weekend I visited Leo Szilard, at this time a patient at Memorial Hospital. Just before Christmas the awful news had reached me that Leo had bladder cancer and that diagnosis might have come too late. Fortunately, by the time I saw him he had taken charge of his radiation therapy and was soon to emerge totally cured. He immediately wanted to gossip about my recent visit to San Diego, while I wanted to talk about our big T2 RNA breakthrough. He said he would take my idea seriously only when messenger RNA molecules were shown to exist in uninfected as well as phage-infected cells. I told him this was to be our next research objective. A young French biochemist from the Institut Pasteur, François Gros, was to come to Harvard for the summer to search for messenger RNA in uninfected *E. coli* cells.

My main goal soon became persuading Matt Meselson to accept Harvard's offer, not Jonas Salk's. His visit to look us over came during a week of fortuitously perfect April weather, a seduction in itself compared with Pasadena smog. Matt, unlike Benzer, quickly said yes to Harvard, telling Paul Doty and me that he would arrive as soon as appropriate research space could be renovated. Mel Cohn, in contrast, opted for the Salk Institute, and Aaron Moscona, to my never hidden delight, decided to remain in Chicago. Also to Harvard's long-term benefit was the acceptance of Keith Porter. In contrast, John Torrey's decision to leave England meant that botany at Harvard would likely continue intellectually vapid.

During that time, Celia Gilbert often invited me for vodka-dominated meals at which one of Wally's young colleagues, the theoretician Sheldon Glashow, was often present. Wally was then twenty-eight years old and had been an assistant professor of physics for two years. Surprisingly, he now found himself more excited by our T2 RNA experiments than by his own attempts at high-level physics. Eagerly he was soon to drive up with Alfred Tissières, François Gros, and me to the 1960 Gordon Conference on Nucleic Acids in New Hampshire, afterward assisting François in his summer pursuit of a T2-like RNA in uninfected bacterial cells. Most conveniently, François and his wife, Françoise, also a scientist, were living in the tiny 10½ Appian Way flat that my father had moved into two years before upon his early retirement from his job in Chicago. Dad was off on a lengthy tour of Europe, having enjoyed a similar trip the year before. To my delight these trips showed that he could now be on his own for long periods, coping with if never quite moving beyond the loss of my mother.

At the Gordon Conference, we learned that Sydney Brenner would soon be going to Caltech to do experiments with Matt Meselson that might independently prove the existence of messenger RNA. In April, François Jacob came to Cambridge to talk with Sydney and Francis Crick and persuaded them too that ribosomes by themselves did not carry genetic instructions for ordering amino acids in protein synthesis. To confirm the hunch, Sydney soon proposed to Meselson that his ultracentrifuge tricks should let them see new T4 mRNA molecules bound to ribosomes made prior to phage infection. By late July, Sydney triumphantly returned to Cambridge with his prediction confirmed. Only later, when Sydney told us about his Caltech results,

Wally Gilbert in my lab, summer 1960

did he learn the details of Bob Risebrough's independent demonstration of mRNA.

That summer of 1960, the lab was graced with several lively Radcliffe students who wanted technician jobs so they could stay around Harvard Square for the summer. Particularly fun to have about was the fetching Franny Beer, the red-haired daughter of the equally red-haired Sam Beer, Harvard's resident expert in American politics. I first knew Franny from my Biology 2 lectures, in which she regularly sat smiling in one of the front rows. Spotting her soon after in the Biolabs, I learned that she was very fond of dogs and keen to be a vet. We both revealed ourselves to each other as loyal Democrats, but at first I couldn't share her enthusiasm for John Kennedy. I still hated his father, Joe, for his past German sympathies. Earlier, in June, Franny and I had watched the Harvard commencement to see her hero march by as an overseer. Like Eleanor Roosevelt, I was then still rooting for Adlai Stevenson to again be the Democratic candidate. Franny followed her father's hopes that John Kennedy, a much stronger potential nominee, would prevail at the convention. That fall Franny, who was mad about rock and roll, brought me to a late gathering of undergraduate rock fans. I was never so out of place, realizing that I should best cherish Franny as a surrogate kid sister.

In the fall, Diana de Vegh was no longer available for Henri IV lunches. John Kennedy's now active campaign for the presidency had put her previous year's studies to be an Arabist into proper perspective. Now she was doing campaign work elsewhere. But into the lab came the Radcliffe senior Nina Gordon, doing an undergraduate project and seeing that I got invited often to her Radcliffe house, where I could feed my hopes of finding a suitable blonde. Increasingly the presidential campaign dominated emotions at Harvard, and I went to watch the debates between Kennedy and Nixon on Alfred and Virginia Tissières's TV set in their big apartment on Sparks Street. They had lived in those spacious early-twentieth-century rooms since their marriage two summers before, and the same apartment would soon be occupied by Matt Meselson and his very new wife, Katherine. Matt and Katherine had met over the summer in Colorado at the Aspen Music Festival, where Katherine was studying the flute.

By early September I was very much a Kennedy partisan, hating Nixon even more and excited that so many Harvard professors were working as Kennedy advisors. Following the public opinion polls with increasing apprehension, I followed the election night cliffhanger with Alfred and Virginia. The Dotys then were part of an older Harvard group close to the Kennedy campaign. Too tired to stay up till Kennedy's victory was ensured, I took comfort in knowing that my mother had long worked for the Democratic political machine, which would not let Kennedy lose in Illinois.

The weeks that followed Kennedy's victory were in no sense anticlimactic. The main question in the air was who from Harvard would be called to be part of the new administration. Arthur Schlesinger's departure to help Kennedy as a speechwriter was virtually taken for granted. Everyone was equally pleased by the selection of John Kenneth Galbraith as ambassador to India and Edwin Reischauer as ambassador to Japan. Most excitement came from McGeorge Bundy's nomination as the president's national security advisor with West Wing offices. For his chief deputy Bundy further raided the Harvard faculty, picking his friend the economist Carl Kaysen.

Before taking office, Kennedy saw fit to resign from Harvard's Board of Overseers, promising to attend its January meeting just prior to his inauguration. For several weeks I anticipated having him listen to me speak, since I had been asked with Frank Westheimer to brief the overseers about new opportunities for research in molecular biology and biochemistry. But our president-elect did not get to Harvard that day, having more pressing matters to attend to. The occasion, however, gave me my last opportunity to speak to McGeorge Bundy as dean. He had raised my curiosity in the weeks before by asking me to come and see him. So I half dreamed that I also might be asked to move to Washington. At the last moment, however, his aide Verna Johnson phoned me to cancel the appointment. Taking me aside at the overseers' meeting, Bundy wanted to personally tell me the good news that I was being promoted to full professor as of July 1. He then mischievously added that no higher academic accolade could ever come my way.

Remembered Lessons

1. Teaching can make your mind move on to big problems

Eminent researchers who revel in trivial or nonexistent teaching loads may be availing themselves of a luxury no thinker can well afford. When I'm not challenged by an immediate need to make sense of incompatible observations, my mind too often runs slowly. A very strong incentive for coming to grips fast with unexplainable experiments is the need to lecture about them. For this the best audiences are advanced undergraduates or graduate students, who know enough to have reactions that may spark a flash of insight. In the early 1970s, when lecturing about DNA duplication in such a fog of uncertainty, I suddenly saw why viral DNA molecules have redundant ends that become linked during their replication processes. The idea that this was a device to copy their ends was too pretty to be wrong.

2. Lectures should not be unidimensionally serious

It is no fun to either give or listen to hourlong talks that provide nonstop flows of dry facts or even ideas. Presentations of all kinds should alternate easy-to-understand and familiar material with the messages that are more difficult to assimilate. At Harvard I tried to put a human face on experiments, adding asides about personalities and letting my listeners put themselves in the place of the experimenter, as eventually they would need to do.

3. Give your students the straight dope

In my Biology 2 lectures in the early 1960s, I regularly gave one titled "Against Embryology," since its main point was that multicellular organisms were best put on the back burner until we understood the basic nature of life by studying single-celled bacteria. The early sixties were not a propitious time, for example, to go to the Marine Biological Laboratory at Woods Hole to study sea urchins. Those who went instead to Cold Spring Harbor to pursue genes within bacteria would have much brighter futures. This was not a message that most of my fellow biology professors agreed with, and many of them thought it inappropriate for me to announce it to my students. But to sugar-coat science that is going nowhere ill prepares students for their futures.

4. Encourage undergraduate research experience

If one or more lab benches were free, I automatically accepted bright undergraduates keen to do research under my supervision. Often they were undecided between medical and graduate school and benefited from seeing the differences between scientific and clinical challenges. Being part of a research group, moreover, let them see that personalities often are as important as brains in pushing forward the scientific frontier. It is also true that a certain kind of aptitude is required to do successful research. You frequently spot individuals in labs whose first-rate talents may never come out through exams. They come alive only when they are challenged with "new unknowns" as opposed to "old knowns."

5. Focus departmental seminars on new science

The quality of a scientific department is generally revealed by its weekly seminars. Star scientists likely will travel only when they see themselves benefiting from being away from their home base. Seminars that fail to attract broad student audiences will likely bore the largely faculty-constituted audiences, there only for reasons of department loyalty. It's best to invite speakers from emerging disciplines not yet established on your campus. Choosing too many speakers from friends of senior faculty risks giving your students no more than what they already have. Younger faculty members, for the most part, should be in charge of arranging and hosting potentially exciting speakers. They have more time and incentive to do this job well, as they anticipate meeting minds that could enrich their future intellectual lives.

6. Join the editorial board of a new journal

Editorial boards of preexisting journals seldom change fast enough to accommodate new scientific disciplines. A new discipline creates a new discourse and requires a new journal. Editors rooted in the past may not know how to assess the importance of new science, or even whom to approach as a referee. Only six years passed between the finding of the double helix and the founding of the *Journal of Molecular Biology.* At first I was hesitant to join its editorial board and spend the time looking for the wheat among the chaff. But when the protein crystallographer John Kendrew became its chief editor, I knew the *JMB* would attract high-quality papers. In return for executing the responsibility to see that important new ideas got out as soon as possible, I was also among the first to benefit from knowing about them.

7. Immediately write up big discoveries

We made a bad mistake in not immediately publishing our lab's February 1960 discovery of T2 messenger RNA. At the time Wally and I wanted the story filled out a bit more by the simultaneous demonstration of *E. coli* messenger RNA. But the latter task proved much trickier than initially expected. Meanwhile, at Cambridge, Sydney Brenner and François Jacob came independently to the concept of messenger RNA in late April, with Sydney soon proving its existence through experiments with Matt Meselson at Caltech. Though we published simultaneously, Sydney let it be known that I had delayed their publication, leading others to believe our Harvard experiments were derivative of theirs. In fact, they predated them by four months.

8. Travel makes your science stronger

No matter how prestigious your own institution, at any given moment the real action in your specialty is likely happening elsewhere. Living in Boston does not mean that you need not continuously monitor the action in other scientific hot spots such as Stanford, Caltech, or La Jolla. Turning down invitations to speak before their audiences works against your future. By moving out of your own turf, you are likely to spot clever graduate students and postdocs who might enhance your own environment. Learning first about clever brains through their publications likely means that someone else has already recruited them. Naturally, there is a point beyond which traveling becomes counterproductive. Whenever possible, you should not cancel lectures for key undergraduate courses. But when you don't have any to give, much time should be spent seeing high-level science done elsewhere.

9. MANNERS NOTICED AS A DISPENSABLE WHITE HOUSE ADVISER

I **WAS** to wait eight months before the Kennedy administration let me know, in September 1961, that my talents might be of use to them. After we had lunched at the long head table of the Faculty Club, Harvard's physical chemist, George Kistiakowsky, motioned me aside to ask whether I would like to assist the President's Science Advisory Committee (PSAC) in evaluating our nation's biological warfare (BW) capabilities. Curious ever since the end of World War II as to what BW weapons we might have developed, I indicated my availability whenever PSAC wanted me. Now some three years old, PSAC had been created by President Eisenhower as a response to the shock of Sputnik's moving the Soviets into space ahead of us. After James Killian, then president of MIT, George had served as its second leader, reflecting Ike's respect for his acumen at applying science to military purposes. At Los Alamos, his long experience with explosives was used in the fabrication of the first nuclear weapons.

PSAC was now headed by Jerome Wiesner of MIT's big Electronics Lab, who at the war's end was also at Los Alamos. Most of its members were physicists and chemists, reflecting a major preoccupation with nuclear weapons and missiles. George was still a member, as was Paul Doty, who was hopeful that with JFK as president we might be able to slow down, if not stop, the testing of ever bigger hydrogen bombs. Soon I filled out several White House forms for an FBI background check necessary to get me a top-secret security clearance. Only at that level of authorization could I get into Fort Detrick, the nation's big,

*Paul Doty (left) with Jerry Weisner (right) and
the president who gave them hope*

rambling biological warfare complex, twenty-five miles to the north of
the D.C. line in the foothills of the Blue Ridge Mountains.

That fall Wally Gilbert was increasingly in the Biolabs, coming over
from the Physics Department, where he still had serious teaching
responsibilities. The messenger RNA concept was on everyone's mind,
with the previous June's Cold Spring Harbor symposium dominated
by its implications. Seeing newly made mRNA molecules functioning
in the *E. coli* cell-free systems made Alfred Tissières wonder whether
RNA molecules containing only single bases might also stimulate protein synthesis. But to his disappointment, the polyadenylic acid, or
poly A (AAA . . .), from Paul Doty's lab had no apparent template
capabilities. Alfred then put synthetic RNA out of his mind until he
came along with Wally and me to hear Marshall Nierenberg's electrify-
ing announcement at the Biological Congress in Moscow in August
that polyuridylic acid, or poly U (UUU . . .), coded for polyphenylala-
nine. Later poly A was also revealed to have template capabilities coding for polylysine. Its mRNA-like activity had been missed by Alfred,
who had the misfortune of being given aggregated poly A incapable of
binding to ribosomes.

Upon his return from Moscow, Wally moved into Alfred's former office lab to study interactions between poly U and ribosomes. Over the coming year, Alfred was to spend periods in Paris and Cambridge waiting for his new lab to be completed in Geneva. Several nights before the Tissières were to depart, they joined me and Franny Beer, again my summer technician, for an elegant private supper at the American Academy's new embassy-like home, Brandegee, southwest of Boston. Franny and I drove out in my MG, which I planned to let her borrow while I was away for the coming six weeks. After Moscow, I was to go on to Cambodia, where my sister's husband, Bob Myers, was our CIA station chief, and then to Japan, where Masayasu Nomura would give me an insider's tour of the country.

That summer Franny was my daily sounding board about my new Radcliffe friends, the striking blond twins Sophia and Thalassa Hencken. Living in a large, comfortable house in posh Chestnut Hill, they existed in the perpetual shadow of their mother, a garden expert with her own TV show. Though seemingly destined to marry a Social Register type, Thalassa had just discovered a handsome young Pakistani engineer who possessed more panache than was usually dealt out to the suitors of future members of Boston's Vincent Club. Sophia, the less flamboyant of the two, had a boyfriend from New Orleans who, though not appropriate for the Brookline Country Club, did a skillful rendition of Gilbert and Sullivan.

The twins' mother was planning a big party for their twenty-first birthday, to be held in their home in mid-October. I had hoped that my multiple letters and postcards from the Royal Hotel in Katmandu would secure me an invitation to be at either Sophia's or Thalassa's side for the big night. Alas, that did not come to pass, although I was invited and Franny graciously came as my date. At the party, the twins were somewhat upstaged by the elegantly tall sophomore Ann Douglas Watson, no relation, whose obvious social and intellectual superiority over the males her age made me wonder hopefully whether the promise of not having to change her name afforded me any advantage as a suitor. But the real catch at the party, all too clear to Mrs. Hencken, was Desmond FitzGerald, the future Knight of Glin, then over from Ireland to study art history at the Fogg Museum.

Soon after, the twins had Desmond invite me to a Saturday night party that he gave at his Massachusetts Avenue flat with Dorothy Dean. Her regal black-draped form often graced the lunch scenes at University Restaurant or Hayes-Bickford's and even more the largely gay evening crowds at Club Casablanca beneath the Brattle Theatre. In talking first to Thalassa, who professed ignorance about most of the other guests, I found my ear tuning into the opinionated, laughing voice of Abby Rockefeller, the youngest of the guests and the eldest daughter of David and Peggy Rockefeller. Instead of going on to college, Abby was studying the cello in Boston, living at a friend's home north of Harvard Square. So with my windowless MG now entrusted for the winter to Miss McCartney's Brattle Square garage, the next afternoon I walked up Brattle Street to the Churchill family residence to continue our spirited conversation of the night before. Over tea we concurred that no more than pennies would ever trickle down from the haves to the have-nots.

By then my security clearance had materialized, and I was soon making regular flights to Washington as part of PSAC's new Limited War Panel. Its recent creation was PSAC's response to the ever growing American involvement in Vietnam. With the use of nuclear weapons ruled out ever since Eisenhower had decided not to so rescue the French at Dien Bien Phu, it was unclear how to keep South Vietnam from falling to the Viet Cong. No one on Bundy's staff thought a massive deployment of ground troops was the answer. Whenever their southern borders were truly threatened, the Chinese could supply more bodies as cannon fodder than any American president dared contemplate matching. Use of highly lethal chemical and biological agents was also a Rubicon the government had no wish to cross. So the army's chemical and biological warfare units were considering deployment of "incapacitating agents" that would put enemy soldiers out of action only temporarily. Secretary of Defense Robert McNamara apparently liked this idea, and PSAC's task was to give JFK an independent appraisal of their possible military feasibility.

The first new chemical agent I was to hear about was in fact a killer—but only of plants. Agent Orange was on the agenda of my first visit to the Executive Office Building, where the southeast side of the third floor, once occupied by Secretary of State Cordell Hull, now contained PSAC's offices. Speaking to the full Limited War Panel, a Green Beret officer explained how spraying this herbicide along roadsides had cut down Viet Cong ambushes. Were his presentation a seminar, I would have questioned his lack of statistical analysis. But as a mere consultant, I thought it prudent to stay quiet at my first briefing by military officers. Later Vince McRae, PSAC's deputy handling limited war matters, let me know he never challenged the competence of officers during military briefings. This was for their superiors to do if they felt so inclined. PSAC's effectiveness on military matters depended upon the Department of Defense seeing the committee as a potential ally for bending the president to their will. Whether Agent Orange reduced ambushes was for the army alone to judge. PSAC's place was to judge whether the chemical's use posed any negative health consequences for the military personnel involved in herbicide spraying. But here again we were assured that such defoliants were no danger to humans.

My conversation with Vince allowed me to find out where my glamorous Radcliffe friend Diana de Vegh was working in the White House. Soon learning that her office was on the floor above, I bounded up the stairs to find her in conversation with her boss, Marcus Raskin, a junior staff member of McGeorge Bundy's National Security Council. Earlier employed by the liberal Democratic congressman Bob Kastenmeier from Wisconsin, Marc was now the Security Council's token left-winger. Having Raskin around, Bundy believed, might afford him more than one type of option for handling a potentially tricky foreign policy dilemma. Much later I learned that Marc's earlier candor about Cuba had by then already put him out of the loop of important decision making. Diana, however, showed no awareness of her office's irrelevance to national security, elated by a helicopter ride over to the Pentagon earlier that day. As she already had plans for the evening, we agreed to have dinner on my next trip to Washington.

Also then a consultant to PSAC was my Harvard colleague E. J. Corey. A first-rate organic chemist, E.J. was to focus on chemical agents, while I handled biological ones. He would go to the Aberdeen Proving Grounds to check up on the Chemical Corps, and I would be calling at Fort Detrick to get the inside skinny on our biological warfare programs. When E.J. and I later put together a report that had the potential to reach JFK, we used E.J.'s ultrasecure safe in the Converse Memorial Laboratory office to store the top-secret materials. Early on, we were briefed on corresponding Soviet efforts. We saw photos, most likely predating Gary Powers's U2 overflights, showing grid lines interpretable as Soviet biological and chemical weapon testing sites. By then the Soviets clearly had the capabilities to deploy deadly organic phosphate nerve toxins in the United States on a mass scale. But would they ever do so if they thought we might respond with a nuclear blast? Moreover, would any serious military establishment take the chance that a shift in wind direction might cause a cloud of nerve gas to drift over friends rather than the target?

Much more urgently in need of serious PSAC review was the Chemical Corps' incapacitant BZ, about which the corps was very enthusiastic. Volunteers exposed to it temporarily became zombie-like without apparent long-term consequences. Might this agent win battles without killing enemy soldiers? But in conditions of extreme heat, would individuals so drugged become fatally dehydrated? Even more worrisome, volunteers exposed to BZ initially experienced delusions reminiscent of those caused by LSD. So neither E.J. nor I saw BZ as a wise measure to neutralize the Viet Cong.

The evening before my first visit to biological warfare headquarters in Fort Detrick, I stopped over in Washington for the deferred supper with Diana de Vegh. We dined at the red-leather-upholstered Jockey Club in the Fairfax Hotel near Dupont Circle. It was *the* place for top executives and politicians to see or be seen, and nobodies were hard to find there during the dinner hours. Diana apparently expected less to see than to be seen, because she did not wear her glasses, without which faces farther away than mine were all a blur. Already part of the Georgetown "New Frontier" crowd, she avoided talk about JFK, focusing on her recent weekend with Secretary of the Treasury Douglas Dillon and his wife, Phyllis.

A power of a very different sort greeted me when I first arrived at the officers' club of Fort Detrick. I was met by the civilian scientific director, the Texas-bred Riley Housewright. Long attached to the biological warfare effort, he had joined the army upon finishing his wartime bacteriology Ph.D. at the University of Chicago. Over lunch Riley told me he viewed his Detrick program as a distasteful national necessity. Afterward, we toured the huge Detrick complex escorted by several army personnel. After being shown a large variety of bomb devices intended to aerosolize biological agents, I was put in protective clothing and taken into a large factory-like building with huge containment facilities for growing and harvesting dangerous pathogens. Then we went back for a briefing on two promising biological incapacitants, Venezuelan equine encephalitis (VEE) and staphylococcus enterotoxin protein.

Of the two, VEE was much further along toward possible tactical use. Though VEE is normally transmitted by mosquitoes, Detrick scientists had shown it also could be transmitted to animal hosts by aerosol mists. Delivered this way, it would likely also infect human beings. Though I was told that adults usually recovered from VEE infection with no long-term brain damage, this high-fever-inducing virus sometimes kills the very young or very old. In my mind employing it in Vietnam, or for that matter anywhere else, should be out of the question. In contrast, I saw much promise in pushing ahead the staph enterotoxin program. While it would make you vomit continually for up to twenty-four hours, ruining church picnics and similar occasions, there were no known fatalities associated with infection. In leaving I told Riley I was surprised that so little was known on our side about the anthrax toxin, despite constant worries about its deadly properties reported in the popular press. Could it easily be weaponized for use against innocent civilian populations without warning?

At that time I was technically on sabbatical leave from Harvard, intending to take up a visiting fellowship at Churchill College in Cambridge. I had not anticipated my PSAC duties when I made plans to see MRC's brand-new Laboratory for Molecular Biology (LMB) in action. I planned to start work there with RNA phages, whose existence had been discovered the year before at Rockefeller University by Norton Zinder and his graduate student Tim Loeb. Until Loeb finished his thesis experiments, Zinder did not want to send out aliquots of his f2 phage to possible competitors. In the year that followed, several additional RNA phages were also discovered, one of which, R17, Sydney Brenner had already got possession of. Now in his lab at the LMB, I wanted to purify R17 as a first step toward studying its relatively small, single-stranded RNA molecule of likely fewer than four thousand nucleotides. They might be super messenger RNA templates for use in in vitro (cell-free) protein synthesis studies. Upon arriving in late March, I was joined by Nina Gordon, who the year before had done senior thesis research in my lab. Now she wanted to be in Europe near her Italian-born theoretical physicist boyfriend, Gino Segre, then in Geneva. Over the next two months, Nina and I were distressingly set back by contamination of our host *E. coli* bacteria by more conventional DNA phages. It was only when I got back to Harvard in early June, where I could enlist the expert hands of my graduate students, that our RNA phage research could effectively start.

While still in England I hoped to learn more about anthrax by visiting the UK's top-secret biological weapons lab at Porton Down, near Salisbury. But its anthrax program proved no more advanced than that at Fort Detrick. In 1942 the UK had conducted extensive anthrax testing on a small island off the coast of Scotland, but the program's current leader, David Henderson, was not inclined to spend any more money on a weapon that he believed could destroy the moral authority of the British government. When temporarily back in Detrick, I was briefed in greater detail about a program on rice blast, a fungal pathogen, about which enterprise I had heard when last there. While geneticists elsewhere were working to develop new rice strains resistant to rice blasts, those at Detrick concentrated on developing appropriate blast strains for destroying the rice crops of North Vietnam. Producing rice blasts in large amounts was within Detrick's capabilities, but delivering them was another matter. Helicopter delivery was deemed totally impractical, and no operational American fighter pilot had the radar capabilities necessary for night spraying missions over the Red River delta rice paddies. Later an air force officer was to fill me in on a still-secret radar-guided, terrain-hugging bomber then being evaluated near Dallas.

Likely it was only a bureaucratic snafu that had let me inside Detrick's Special Project facility, a part of the operation I would never see again. There scientists worked not for the Department of Defense but for the CIA on poisons to be used for assassinations. Among others they were keen to employ was the puffer fish toxin. Synthesizing it, however, was a chemical challenge worthy of the best organic chemists, and they had approached Harvard's Bob Woodward for help. Somewhere within this nondescript, drab building was likely stored the chemical agent that Bobby Kennedy later hoped could be slipped to Fidel Castro.

Many faculty members at Harvard who were New Frontier boosters were embarrassed when JFK's thirty-year-old brother Edward Moore (Teddy) Kennedy campaigned against the incumbent George Lodge to become the new junior senator from Massachusetts. His one year of experience as an assistant district attorney was presumptuously paltry. And Teddy's undergraduate years at Harvard were tainted by the scandal of his having sent someone else to take a language exam in his place. Though he later obtained a law degree from the University of Virginia and passed the bar in 1959, Harvard was not proud to count him as one of its alumni. Also running for senator, as an independent, was Harvard history professor Stuart Hughes, whose campaign was based largely on opposition to the nuclear weapons race. Sam Beer, Franny Beer's political scientist father, did not warm to Hughes, believing that he was impractical and unelectable and that as Democrats we should be backing someone who would strengthen JFK's support in the Senate. Soon after I was invited to see Teddy in action at a gathering Sam was holding for important Harvard colleagues at the Hotel Continental in Cambridge. That day Teddy was clearly less impressive than his appealing, fair-haired wife, Joan.

PSAC's oversight of poisons took on more humane considerations after Jerry Wiesner read Rachel Carson's *Silent Spring*, serialized by the *New Yorker* in June of 1962. Carson argued that chemical pesticides were fast spreading through the world's food chains, posing an immediate threat to the global environment. Not only were they killing off fish and songbirds, they possibly threatened human existence. With her thesis quickly generating a firestorm of public concern, JFK himself was drawn into the controversy and stated that Carson's book had led his administration to take the pesticide threat seriously. No federal agency then had a real mandate for an honest investigation of the chemicals' ecological consequences. The obvious candidate, the U.S. Department of Agriculture (USDA), was too cozy with the agricultural chemical industry. So Jerry assigned his Life Sciences deputy, Colin MacLeod, to head up a special PSAC panel, which Paul Doty and I were asked to serve on. Meeting first on October 1, our deliberations momentarily came to a halt when the Cuban missile crisis drew PSAC's attention elsewhere. Only years later did I learn that one of the nation's contingent responses that scary week was to have jets from Homestead Air Force Base, south of Miami, drop VEE-filled devices on Cuba.

Our panel dealt with two major groups of pesticides: the long-lived chlorinated hydrocarbons, of which DDT was the best-known, and the much more toxic, short-lived organic phosphates, such as Sevin. The latter originally were developed as nerve agents for military deployment but later synthesized as less toxic derivatives, such as parathion, to kill insects. The use of both pesticide groups was steadily increasing, with many insects in turn developing genetic resistance, especially to the chlorinated hydrocarbons. Because of their much greater stability, Carson had focused more attention on the chlorine-containing pesticides, pointing out ever-increasing concentrations in the fatty tissues of creatures throughout the food chain. While large amounts of DDT given to human volunteers had no short-term effects, its more toxic derivatives, such as dieldrin, might well pose public health threats. An already widely used pesticide, dieldrin was a nasty liver toxin at high doses. More worrisome, mice exposed to it at much lower levels were developing liver adenomas that conceivably might develop into malignant carcinomas. But with the Federal Drug Administration (FDA) calling these adenomas benign, the USDA blocked the invocation of the so-called Delaney amendment, which prohibited cancer-causing agents in the nation's food. If the FDA were to ban outright all chlorinated hydrocarbon pesticides, however, American agriculture would have been deprived of a chemical that had become vital to its productivity. Prudence suggested that the proper course was to recommend sharp curtailment of dieldrin use until the question of its carcinogenicity was settled.

Only after a thorough review of how the USDA and FDA dealt with pesticides did our panel invite Rachel Carson to appear. Pleased at being asked, she was nevertheless perfectly even-tempered on that late January day, giving no indication of the nutty hysterical naturalist that agricultural and chemical lobbyists had portrayed her to be. The chemical giant Monsanto had distributed five thousand copies of a brochure parodying *Silent Spring* entitled "The Desolate Years," describing a pesticide-free world devastated by famine, disease, and insects. The attack was mirrored in *Time* magazine's review of *Silent Spring* deploring Carson's oversimplification and downright inaccuracy. Two weeks after meeting with her, our panel finished a much debated first draft of our presidential report. Though it accepted as indispensable the role of pesticides in modern agriculture and public health (e.g., to control mosquitoes), most of it was devoted to dangers that pesticides posed for human beings, fish, wildlife, and the environment.

The USDA reacted to the draft with instant fury, and Secretary Orville Freeman wrote to PSAC that in its present form the report would profoundly damage U.S. agriculture. After more pages on the benefits of pesticides had been added and the full PSAC panel had approved it, the USDA then demanded that they make a full review of it before the president released it. But Jerry Wiesner held firm, refusing to add a blanket statement that the food of our nation was safe or to remove the final sentence, which paid tribute to Rachel Carson for alerting the public to the problem. To our great relief President Kennedy released the document uncorrupted on May 15, 1963.

By then Diana de Vegh was no longer part of Marc Raskin's attic office above PSAC in the Executive Office Building. Despite having recently purchased a house on a quiet street near Georgetown University, she had precipitously left for Paris. My Washington meals increasingly had to be taken with fellow panel members or with Leo Szilard and his wife, then living out of suitcases at the Dupont Plaza Hotel.

During the Cuban missile crisis, Leo was so terrified that war was about to break out that he left New York for Geneva via Rome, where he tried unsuccessfully to get the pope's attention. A month later, the Szilards somewhat sheepishly returned to Washington, where Leo continued to devise unorthodox schemes to reduce the probability of nuclear war. Now aboveground test blasts were again occurring, with the Soviets breaking the international moratorium soon after the Berlin Wall went up. Six months later, our bomb makers were to follow suit.

At that time, my major political concern was ever-expanding U.S. involvement in Vietnam. Then just back in Washington were my sister, Betty, and her husband, Bob, the former CIA station chief in Cambodia. Bob's more than fifteen years of experience in the Far East had convinced him that sending more American troops to Vietnam would create a quagmire that would long haunt our nation. But he knew that many U.S. Army officials were more optimistic. In their ranch-style house, within easy commuting distance to CIA headquarters in Langley, Virginia, I expressed my belief that our Limited War Panel had nothing to offer the American cause in Vietnam. By then I had learned that staph enterotoxin could no longer be considered merely an "incapacitating" agent. Monkeys exposed to it through their lungs promptly died. We had to assume that humans would suffer the same fate if exposed. I informed John Richardson of this fact four months later, when he came out to Bob and Betty's house following his removal as station chief in Saigon, an action taken to mean that the United States no longer supported the corrupt Diem government. Two years before, on my way to Cambodia for a family visit, I had been warned by Bob Bloom, then leading the CIA's secretly funded Asia Foundation, that any successor to Diem was likely to be even worse.

A month later I was with the "superspook," Desmond FitzGerald, whose house I came to one mid-June evening to take his stepdaughter, a classmate of Abby Rockefeller's, to dinner. A member of the Social Register elite that helped found the CIA, Desmond knew from his experience in the Philippines that bribes, not soldiers, were generally the best way to promote American foreign policy objectives in Asia. His mind seemed elsewhere when I indicated doubt that Fort Detrick's rice blast arsenal could prevent North Vietnam from continuing to support the Viet Cong. Only twenty-five years later did I learn that Desmond had been entrusted by Bobby Kennedy with the task of assassinating Fidel Castro.

Though I attended a full Limited War Panel meeting in early May, my main PSAC role then was to lead its new panel on cotton insects. Its origin lay in a request to JFK from Arizona's Senator Carl Hayden that the federal government somehow prevent the boll weevil from spreading from Mexico into his state's highly profitable irrigated cotton fields. Pesticides already amounted to 20 percent of total production costs in southeastern states, and every year boll weevils were becoming increasingly resistant to the chlorinated hydrocarbons being used. When our panel first met in nearby Beltsville, Maryland, we were briefed about the problem as well as a possible solution. The sterilization procedures that had supposedly eliminated screwworms from selected horse-racing regions of Florida promised a theoretically ideal method of purging the cotton crop of the boll weevil. But the technique's application to weevil eradication seemed practically daunting. Producing and releasing enough sterile boll weevils to significantly reduce their population could easily cost several billion dollars.

During our subsequent tours of the cotton fields of Mississippi, Texas, California, and Mexico, the entomologists who dominated our panel believed integrated pest management approaches could save farmers from further pesticide expenditures. Our first stop was the Boll Weevil Research Laboratory at Mississippi State College in Starkville, not surprisingly located in the congressional district of the powerful Jamie Whitten. As chairman of the House Appropriations Committee, Whitten was more influential in shaping agricultural research policies than the secretary of agriculture. Later as we drove toward the vast cotton fields of the still British-owned Delta and Pine cotton plantation near Greenville, the homes of the "hoe-hands" who weeded the cotton fields struck me as even more run-down than the dwellings bordering the rice paddies of Cambodia that I had visited two years earlier. Later the Delta and Pine official showing us around his vast domain remarked that his farm laborers lazily stopped hoeing every time his car passed out of sight.

On our way to the USDA cotton insect lab near Brownsville, Texas, we were lucky not to be in a convertible when our car was doused with pesticides released by a small plane flying overhead. Pesticide advertisements were ubiquitous on the large roadside billboards, where competing agrochemical companies touted the merits of their respective pesticide brews like so many aftershaves. Before I got to Texas, I had hopes that the boll worm, in some years a worse pest than the boll weevil, might be best controlled by spraying with polyhedral viruses. Similar viruses had already been used to control the spruce bud worm in Canada, but such an enterprise could not be supported here, as we found to our dismay that 85 percent of the Brownsville laboratory's budget went to salaries. A major function of most USDA regional labs was then to provide patronage jobs for friends of local congressmen. At the Boll Weevil Research Lab, for example, the chief administrative assistant was a close relative of Congressman Whitten.

In the fall we came together three times to hammer out details of our final report. I wrote the introductory sentence: "The boll weevil is almost a national institution." Secretly I hoped that JFK himself might read it and mark me out as a potential speech writer. On the first day of our last scheduled meeting, we were interrupted by Colin MacLeod's deputy, Jim Hartgering, bursting in to tell us that the president had been shot in Dallas. Halfheartedly we tried to refocus on cotton insects until news reached us an hour later that JFK had died. In a state of shock, I walked about the PSAC offices, soon drifting upstairs to see Marc Raskin, who for months had wanted to resign from his sideline position on Bundy's National Security Council to start his own foreign policy institute. We wondered under what circumstances Diana de Vegh would hear the news. That Lyndon Johnson was to be our president was at that moment emotionally impossible to accept.

At last I saw no point in hanging around and went back to the Dupont Plaza Hotel, where I was staying to be near the Szilards, rather than the Hay-Adams Hotel, across Lafayette Park from the White House, which was inexpensive in those days and where I had stayed on earlier Washington trips. Always looking forward to his next meal, Leo insisted that Trudy and I quickly go with him to the Rathskeller, across Dupont Circle down Connecticut Avenue. There he obsessed about how one might get Lyndon Johnson to end the nuclear arms race. I didn't have the heart to stay in town and see the funeral cortege that would soon be making its solemn way down Pennsylvania Avenue. Abjectly I flew back to Boston the next morning.

Though Bundy stayed on as national security advisor, Jerry Wiesner soon resigned to return to MIT as dean of science. Almost eight months were to pass before our cotton insects report finally was released in a gutted form. Gone were our recommendations to spend more on cotton research facilities and supplies and less on salaries. Unless many more entomologists were trained to help bring savvier approaches to the fields, we saw no chance of American cotton's escaping its total dependence upon pesticides. But we were told that the new president didn't want us to recommend policies requiring more money for cotton insect research. With our final report likely to have an impact on no one, I saw no reason to oppose its new, more pedestrian opening sentence, "Cotton is the largest cash crop in the United States."

My last day as a $50-a-day PSAC consultant occurred when the Biological and Chemical Warfare subpanel was brought together to evaluate a proposed release of several infectious agents over the Pacific Ocean southwest of the Hawaiian Islands to test whether they would infect endemic Pacific birds. If no such infections occurred, VEE, for example, would finally get a true green light for appropriate military use. When I saw that a lieutenant general had come to preside over the briefing, I knew the army strongly wanted these tests to take place. Already they had co-opted the Smithsonian Institution for ornithological help. That morning I was the only panel member in opposition, in particular arguing that VEE was not an "incapacitating" agent. It killed the very young and very old and should never be sprayed over any civilian-populated areas. Talking later to Vincent McRae, I got the distinct impression that the lieutenant general had wanted a unanimous vote in favor of his Pacific tests. So I was not surprised never again to be called back to the Executive Office Building.

More than a year later, in early June 1965, I was invited to a reception held for Presidential Scholars—those honored as the cream of the nation's graduating high school seniors, under a program LBJ invented—on the south lawn of the White House. I found myself next to the ice skater Peggy Fleming, who in turn was next to General William Westmoreland. Upon the dais Lucy Baines Johnson talked about how we should now strongly support our American soldiers, no longer in Vietnam merely as "observers" but now in frighteningly larger numbers as combat troops. Later going through the reception line, I watched Senator J. William Fulbright attempt civility when briefly speaking to the president. Equally gracious then was Lady Bird Johnson, leading some of us inside to see the executive mansion's reception rooms. I realized at that moment that an era had passed and that seeing the inside of the White House was a now-or-never opportunity.

Remembered Lessons

1. Exaggerations do not void basic truths

Books, like plays or movies, succeed best when they exaggerate the truth. In communicating scientific fact to the nonspecialist, there is a huge difference between simplifying for effect and misleading. The issues that scientists must explain to society—then DDT contamination, today global warming or stem cell technology, say—require far too many years of training for most people to take hold of them in all their nuances. Scientists will necessarily exaggerate but are ethically obliged to society to exaggerate responsibly. In writing my textbooks I realized that emphasizing exceptions to simple truths was counterproductive and that use of qualifying terms such as *probably* or *possibly* was not the way to get ideas across initially. So while some of Rachel Carson's facts have proved less solidly grounded than she first believed, the truth is that man-made pesticides were spreading through the food chains so fast that they were very likely to reach levels dangerous to humans. No good purpose other than the bottom line of the chemical industry would have been served hedging that fact.

2. The military is interested in what scientists know, not what they think

PSAC's briefing by Fort Detrick's staff focused on whether proposed biological warfare agents would be effective if deployed by either our military or the Soviets'. Whether these weapons *should* be deployed was not open for discussion. And so the question as to whether VEE should then be seen as a tactical or strategic weapon was never brought before us. I naively assumed that no one would seriously consider using it in any capacity in the near future, but what may seem absurd to a civilian can be perfectly plausible in a world where options are rarely taken entirely off the table. It is hardly surprising we were never told that the VEE was almost ready for military deployment. We would have gone instantly to McGeorge Bundy, if not the president, to let him know of our opposition to its use at any time. Whether either Bundy or JFK knew how advanced the nation's VEE program was I still don't know. My guess is that they knew no more than our PSAC panel. Top-secret clearance should never be confused with "need to know." I was granted the former but only through my natural curiosity about a building with no apparent function did I learn that one of Fort Detrick's better-funded missions was to advance CIA assassination possibilities.

4. Don't back schemes that demand miracles

Ridding our southern states of the boll weevil by exposing female weevils to irradiated sterile males was a proposal that instantly smelled of nonsense to us experts. No one who briefed us was prepared to say how much it might cost. Even worse, almost all the small pilot tests done to date had failed, with their proponents now saying more research was needed. The sterile male project had an interest in preserving the congressional perception that the Boll Weevil Research Lab was on the verge of something big. Congressman Jamie Whitten could then bask in its supposed glory. Those reading our report knew that we thought the local research was going nowhere, but ultimately it is possible to ignore what even the government's own scientific advisers think. Never mind that producing enough sterile males to blanket the nation's cotton-growing regions might cost more than the profit from an average year's crop.

5. Controversial recommendations require political backing

Our PSAC panel's conclusion that pesticides pose a threat to the environment reached the public only through its release by President Kennedy. If he had owed a major debt to the chemical industry, his staff might have seen to it that passages damaging to those interests were toned down, leaving open the question of whether *Silent Spring*'s argument had merit, and dampening the demand for corrective action. Happily, JFK owed no such political debt, and no White House pressure ever came to bear on us. In contrast, President Johnson's staff saw political harm in a White House report that said the nation's cotton farmers needed more than heavy pesticide spraying to keep their fields financially viable. When our badly gutted cotton insect report came out, most panel members realized we had toiled to no useful end.

INDIVIDUALS nominated for Nobel Prizes are not supposed to know their names have been put forward. The Swedish Academy, which judges candidates and awards the prize, makes this policy very explicit on their nomination forms. Jacques Monod, however, could not keep secret from Francis Crick that a member of the Karolinska Institutet in Stockholm had asked him to nominate us in January for the 1962 Nobel Prize in Physiology or Medicine. In turn, Francis, when visiting Harvard that February to give a lecture, let the cat out of the bag at a Chinese restaurant where we were having supper. But he told me we should say nothing to anyone, lest it get back to Sweden.

That we might someday get the Nobel Prize for finding the double helix had been bruited about ever since our discovery. Just before my mother died in 1957, she was told by Charles Huggins, then the University of Chicago's best-known physician-scientist, that I was certain to be so honored. Though many were initially skeptical that DNA replication involved strand separation, this doubting chatter went silent after the 1958 Meselson-Stahl experiment demonstrated that very phenomenon. Certainly the Swedish Academy had no doubt as to the correctness of the double helix when they awarded Arthur Kornberg half of the 1959 Physiology or Medicine prize for experiments demonstrating enzymatic synthesis of DNA. When photographed shortly after learning of his Nobel, a beaming Kornberg held a copy of our demonstration DNA model in his hands.

As the October 18 date for announcing the year's Nobel in Physiology or Medicine approached, I was naturally jittery. Conceivably the responsible Swedish professors had requested more than one nomination, reflecting split opinions during preliminary caucusing. Nonetheless, as I went to bed the night before the prize announcement, I couldn't help fantasizing about being awakened by an early morning phone call from Sweden. Instead a nasty cold I'd caught awakened me prematurely, and I was depressed to realize at once that no word had come from Stockholm. I remained shivering under my electric blanket, not wanting to get up when the telephone rang at 8:15 A.M. Rushing into the next room, I happily heard a Swedish newspaper reporter's voice tell me that Francis Crick, Maurice Wilkins, and I had won the Nobel Prize for Physiology or Medicine. Asked how I felt, all I could say was, "Wonderful!"

First I phoned Dad and then my sister, inviting each to accompany me to Stockholm. Soon after, my telephone began to buzz with congratulatory messages from friends who had already heard the news on the morning broadcasts. There were also calls coming from reporters, but I told them to try me at Harvard after I'd given my morning virus class. I felt no need to rush through breakfast with Dad, so the class hour was almost half over when I walked in to find an overflowing crowd of students and friends anticipating my arrival. The words *Dr. Watson has just won the Nobel Prize* were on the blackboard.

The crowd clearly did not want a virus lecture, so I spoke about feeling the same elation when we first saw how base pairs fitted so perfectly into a DNA double helix, and how pleased I was that Maurice Wilkins was sharing the prize. It was his crystalline A-form X-ray photograph that had told us there was a highly regular DNA structure out there to find. If Linus Pauling's ill-conceived structure had not gotten Francis and me back into the DNA game, Maurice, keen to resume work on DNA the moment Rosalind Franklin moved over to Birkbeck College, might by himself have been the first to see the double helix. He was temporarily in the States when the prize story broke, and held his press conference next to a big DNA model at the Sloan-Kettering Institute. The long-standing rule that a Nobel Prize can be shared by at most three individuals would have created an awkward if not insolvable dilemma had Rosalind Franklin still been alive. But having been

Celebrating my big news with Wally Gilbert (left) and Matt Meselson (right)

tragically diagnosed with ovarian cancer less than four years after the double helix was found, she'd died in the spring of 1958.

After class ended, I soon found myself with a champagne glass in hand and talking to reporters from the Associated Press, United Press International, the *Boston Globe,* and *Boston Traveler.* Their stories were picked up by most papers across the country, clippings from which came to me through the Harvard news office. Often they were accompanied by AP photos showing me in front of my class or holding the hand-size demonstration model of the double helix built at the Cavendish back in 1953. Able to afford the luxury of modesty, I tried to downplay potential practical applications, saying that a cure for cancer was not an obvious consequence of our work. And with my stuffy head and hoarse voice quite apparent, I emphasized that we had not done away with the common cold. This became the quotation of the day in the October 19 *New York Times.* When asked how I would spend the money, I said possibly on a house and most certainly not on hobbies such as stamp collecting. To the question as to whether our work

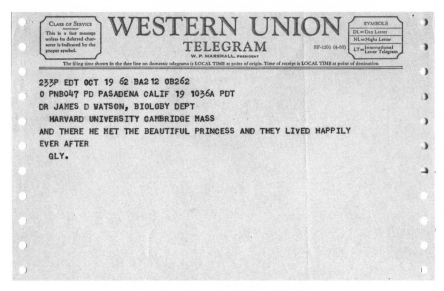

Richard Feynman's congratulatory telegram,
signed with his RNA Tie Club code name, GLY

might lead to genetically improving humans, I answered, "If you want to have an intelligent child, you should have an intelligent wife."

A number of the next day's articles described me as a boyish-looking bachelor whom friends found lively and kindly. Not surprisingly, some reports had bad gaffes such as Maurice's picture above my name, while others reported that I had worked during the war on the Manhattan Project, again confusing me with Wilkins, who had come to the United States in 1943 to work on uranium isotope separation at Berkeley. Naturally the Indiana papers played up my IU background, with the *Indiana Daily Student* quoting Tracy Sonneborn that I was a formidable reader with no tolerance for stupidity and a great respect for smarts. In a similar vein, the *Chicago Tribune* took pride in reporting my Chicago upbringing and appearance on *Quiz Kids,* quoting my father that it was through my childhood interest in birds that I got into science.

A hastily arranged evening blast at Paul and Helga Doty's Kirkland Place house allowed my Cambridge friends to toast my good fortune. Earlier I had talked by phone to Francis Crick, no less elated in the other Cambridge. Most of some eighty congratulatory telegrams arrived over the next two days, while the next week brought some two hundred letters I would eventually have to acknowledge. Joshua Lederberg, whose Nobel Prize had come three years earlier and who'd lived through the pandemonium that goes with it, advised me to follow his trick of replying with postcards from Stockholm. As he was then briefly hospitalized, Lawrence Bragg, our old boss at the Cavendish, had his secretary write of his delight. Unique in addressing me as "Mr. Watson," President Pusey wrote: "It seems almost superfluous to add my congratulations to the many friendly messages you will be receiving." And I had to wonder whether I had been mistaken in backing Harvard's Stuart Hughes for the Senate when not he but Edward M. Kennedy took the time to write, "Your contribution is one of the most exciting scientific achievements of our time."

There were also the inevitable letters expressing not congratulations but the writer's personal hobbyhorse. One from a Palm Beach man, for instance, declared that marriages between cousins are the cause of all the great evils that have afflicted mankind. Here I thought better of writing back to ask whether there had been any such marriages among his ancestors. Several days later, the managing director of the Swedish company that manufactured Läkerol throat pastilles wrote to say that he was having an export carton sent direct from his New York distributor. He noted his product's absence of harmful ingredients, which feature made it suitable for everyday use even in perfect health. Soon I began popping several of his pastilles each day after my cold turned into a sore throat. Unfortunately, they were of no effect, my throat misery persisting through days of celebration.

From a Warsaw, Indiana, podiatrist came advice that all disease results from two simple but pervasive problems—fatigue and respiratory imbalance. Through his research he had learned that these two root pathologies were themselves founded in abnormal foot mechanics and gait. He had treated many people who never caught cold because of therapeutic restoration of normal walking mechanics. Among the beneficiaries of the treatment course, which typically ran three to four years, was the correspondent himself. Another revolutionary way to stop colds was suggested by a New Mexico man of Scottish descent who had noted that the main characters in John Buchan's novels *The Thirty-Nine Steps* and *John Macnab* never even sniffled. This immunity the New Mexico man attributed to strenuous workdays in the cold, fog, and rain.

More poignant, but strange nonetheless, was a letter from a seventeen-year-old Samoan girl in Pago Pago who, after thanking the Lord for his love and kindness, introduced herself as Vaisima T. W. Watson. She hoped that I was related to her father, Thomas Willis Watson, a U.S. Marine supply sergeant during World War II. Her mother had never heard from him following his return to the States. In my reply, I pointed out that Watson was a common name, with hundreds of entries in the Boston phone book alone.

Soon the itinerary of my forthcoming Nobel week, in broad outline at least, was sent to me from Stockholm. I would be housed in the Grand Hotel with my guests. My personal expenses there would be paid by the Nobel Foundation, which would also cover the food and lodging of a wife and any children. As getting there would be my responsibility, the Nobel Foundation would advance some of my prize money for airfare. The prize presentations were to take place at the Stockholm concert hall according to custom on December 10, the date on which Alfred Nobel died in San Remo, Italy, in 1896 at the age of sixty-three. I was expected to arrive several days earlier for two receptions, the first given by the Karolinska Institutet for its winners, and the second by the Nobel Foundation for all laureates except for those in Peace, who always receive their prizes in Oslo from the Norwegian king. At the prize ceremony and at the banquet the following night at the Royal Palace, I was to be dressed in white tie and tails. Meeting me at the airport was to be a junior member of the Foreign Ministry, who would accompany me to all official functions and see me off at my departure.

Making it an even more meaningful occasion was the awarding of the year's chemistry prize to John Kendrew and Max Perutz for their respective elucidations of the three-dimensional structures of the proteins myoglobin and hemoglobin. Never before in Nobel history had one year's prizes in biology and chemistry gone to scientists working in the same university laboratory. The announcement of John and Max's prize came several days after ours was announced, on the same day the physics prize was awarded to the Russian theoretical physicist Lev Landau for his pioneering studies on liquid helium. Unfortunately, because of a ghastly automobile crash that had recently caused him severe brain damage, he would not be joining us in Stockholm. After the double helix was found, the Russian-born physicist George Gamow had much raised my ego by saying that I reminded him of the young Landau. Last to be announced was the literature prize, awarded to the novelist John Steinbeck, who was to deliver his Nobel address at the large banquet in the Stockholm city hall following the prize ceremonies.

In the long letter detailing how to prepare for Stockholm, I was informed that appropriate accommodations would also be reserved for special research assistants. If I could tactfully bring along my lab assistant of the past summer, the Radcliffe junior Pat Collinge, the occasion would be even more decorative. Her fey urchin manners, together with her intense, catlike blue eyes, were likely to have no equal in Stockholm. Alas, she now had a Harvard undergraduate boyfriend of literary aspirations, whom I was unlikely to supplant. Pat promised, however, to help me master the waltz steps that I would need for the customary first dance following John Steinbeck's speech.

For several days, I eagerly anticipated a November 1 state dinner at the White House to which I received a last-minute invitation. Though the event was intended to honor the grand duchess of Luxembourg, I was more keen to see America's royal couple in action. Only six months had passed since they had elegantly honored the nation's 1961 Nobel Prize winners, so I now thought the occasion might find me seated beside Jackie. All such thoughts, however, were abruptly interrupted by the Cuban missile crisis. JFK's speech to the nation on Monday, October 20, was not one to be listened to alone. Nervously, I went to the Doty home to watch it on their relatively big TV screen. Even before the speech was over I knew the gravity of the situation was such that a politically unnecessary state dinner was bound to be cancelled. From then on the president's attention would necessarily be focused on whether the Soviets would challenge the American blockade of Cuba, in which case the prospect of nuclear war seemed very real indeed.

Over the next several days, I had to wonder whether a month hence

I would, in fact, be going to Stockholm. The Soviets might very well set up their own blockade of Berlin. Happily, less than a week passed before Nikita Khrushchev backed down. By then it was too late to reschedule the dinner for the grand duchess. The White House, however, kept me in view and invited me to a December luncheon for the president of Chile. But the thrill that came of seeing the White House envelope vanished when I opened it and saw that the date overlapped with Nobel week. I continued hoping that there might be a place for me at still another White House affair. But by the new calendar year I was no longer a celebrity of the moment.

A postcard from Berkeley brought me some unexpected happiness. It was from my former Radcliffe friend Fifi Morris, who five years before had become quite angry with me over a fondness I developed for her roommate. Equally unexpected was a letter originating in Chicago from Margot Schutt, whose acquaintance I made aboard a ship headed back to England in late August 1953. After graduating from Vassar, she had moved to Boston at the same time I came to Harvard. We had briefly dated in early 1957, until one day when she abruptly hinted that I was not a good enough reason for her to remain in Boston by announcing she would be setting off for London. Now she lived in Chicago and had read that I was scheduled soon to visit the University of Chicago.

The Chicago trip had been arranged some months before the Nobel Prize announcement. Suddenly it became a media event, with visits to my former grammar school and high school hastily arranged. Also making a return visit to Horace Mann Grammar School that day was Greta Brown, the principal when I was there between the ages of five and thirteen. Earlier she had penned me a warm letter recalling my bird-watching days and regretting that my very well-liked mother had not lived to enjoy my triumph. The school auditorium was crammed as I spoke from the stage, gazing once again upon its handsome big WPA murals. The next day, in the *Chicago Daily News* a nearly whole-page spread was headlined "The Return of a Hero" and quoted a teacher remembering me as "very short in stature but with a very eager mind." Later at South Shore High School I spoke to an even larger audience including my former biology teacher, Dorothy Lee, who much encouraged me during my sophomore year.

At the University of Chicago, my new fame caused my scheduled lecture to be relocated to the large law school auditorium. Later I went for dinner to the Hyde Park home of a friend from my phage past, the University of Chicago biochemist Lloyd Kosloff. The next afternoon, I was at the ABC television studios for an afternoon taping of Irv Kupcinet's show. A real Chicago celebrity owing to his *Sun-Times* daily gossip column, Kup also hosted a three-hour talk show on which that day I shared billing with the authors Leo Rosten and Vance Packard.

That evening my dinner with Margot Schutt ended with a surprise twist. At a French restaurant on the North Side, we brought each other up to date on our recent pasts. She was currently working at the Art Institute, where the Sunday afternoon public lectures of my uncle Dudley Crafts Watson had been long appreciated. Though what had always most attracted me about Margot were her almost Jamesian mannerisms, I had not anticipated a Jamesian end to the evening. Over dessert she suddenly asked me if I would have actually gone through with marrying her if she had accepted my advances that spring of '57. I didn't know what to say and the taxi ride back to her flat passed largely in an awkward silence.

The next day I flew on to San Francisco and went down to Stanford to talk science. Then I traveled across the bay to Berkeley, where I stayed with Don and Bonnie Glaser. Two years before, Don had won the Nobel Prize in Physics for his invention of the bubble chamber, causing them to advance their wedding date so that they could fly off together as man and wife to Sweden. In her note of congratulations, Bonnie encouraged me to set my sights on a Swedish princess, suggesting Désirée for both her poise and beauty, as well as having more to say than her two older sisters. So I told them about a telegram from my Caltech friend, the physicist Dick Feynman, wherein he proposed the same scenario with even more irony: "And there he met the beautiful princess and they lived happily ever after." More seriously we gossiped about my Radcliffe friend Fifi, now a Berkeley graduate student. I picked her up the next evening at her flat off Channing Way, driving to Spenler's fish restaurant near the water at the foot of University Avenue. There I again saw the good-natured, intelligent personality the memory of which had led me to conclude she would be a most appropriate Stockholm consort. But still haunted by Margot's Jamesian question, I kept our dinner talk light, letting her know that my father and sister happily anticipated being with me in Stockholm. The evening ended with our deciding to see each other again during my planned February visit back to the Bay Area.

My forthcoming Nobel address soon preoccupied me at Harvard. Maurice was to give his talk on his King's College lab work confirming the double helix; Francis would focus on the genetic code; and I would talk about the involvement of RNA in protein synthesis. Happily, my Harvard science of the past five years was equal to a Nobel lecture. I gave it a dry run in late November at Rockefeller University, using the occasion to visit the BBC offices in New York to let them tape my recollections of the man Francis was before his great talents had become widely appreciated. By then I had bought the necessary white-tie outfit at the Cambridge branch of J. Press, whose first shop in New Haven had long been purveyor par excellence of preppy clothing to Yale's undergraduates. Soon after coming to Harvard, I had begun getting my suits at their Mt. Auburn Street store, finding their clothes to be among the few available that fit my still-skinny frame. Perhaps sensing my high spirits, the salesman easily persuaded me also to purchase for the august occasion a black cloth coat with a fur collar.

Early on the afternoon of December 4 my sister joined Dad and me in New York for our Scandinavian Airlines flight. Our plans were to stop over for two nights in Copenhagen to see friends Betty and I had made when we lived there in the early 1950s. But after crossing the Atlantic, the pilots discovered that Copenhagen was fogged in. So we found ourselves in Stockholm two days earlier than expected. Bypassing customs as if we were a diplomatic delegation, we were whisked by limousine to the storied Grand Hotel, built in 1874, across from the Royal Palace, onto which looked my room, among the finest in the house. I soon joined my sister and Dad for a herring-heavy smorgasbord, where we lunched with Kai Falkman, the young Swedish diplomat who would accompany us to all our Nobel week engagements.

My sister, Betty, my father, and I upon our
arrival in Stockholm in December 1962

After a much-needed nap, we all supped together in the rathskeller of a restaurant in the Old Town, whose buildings date back to the fifteenth century. There Kai told us that the youngest of the four Swedish princesses, Christina, wanted to spend a year at an American university, possibly Harvard, following graduation from her Swedish high school. Conceivably she would like to talk with me during my Nobel visit. Naturally, I pledged to make myself obligingly available to explain Radcliffe's unique relation to Harvard.

Not waking until almost noon the next day, I saw my picture on the front page of Stockholm's *Svenska Dagbladet* together with a chatty article mentioning my interests in politics as well as birds. Unfortunately, the sore throat that had followed my bad cold of October soon forced me to seek medical attention. So my first view of the Karolinska Institutet complex was of its main hospital, not of its research labs. There I was examined by its leading nose and throat specialist, who

My sister, Betty, watches as Princesses Désirée,
Margaretha, and Christina take their seats.

saw nothing worrisome but did reveal he had been a member of the committee that had chosen me for the Nobel. In that official capacity, he greeted me when I arrived at Nobel House the next evening for the reception given by the Karolinska Institutet. This was not a formal event, and I arrived in the pinstripes that I had worn on the air journey to Sweden. The Foundation House also served as the home of Nils Stahle, its executive director, and I was pleased that his pretty, red-haired, unmarried daughter Marlin was at home for the evening, as was Helen, the very blond daughter of Sten Friberg, the rector of Karolinska.

Both girls were later welcome sights at the first formal event of the week, the Nobel Foundation's reception for all the year's laureates. In the grand library of the Swedish Academy, the dominant figure was John Steinbeck, who had arrived in Sweden only that morning. Though his anticipation of the honor had been keen, he was more nervous than happy, worrying about his Nobel address the next evening. William Faulkner's address of 1950 was still remembered with reverence, and Steinbeck was feeling the pressure of expectations.

The Nobel ceremony, December 1962. From left to right: Maurice Wilkins,
Max Perutz, Francis Crick, John Steinbeck, me, and John Kendrew

Francis and Odile Crick
brought their own princesses
to Stockholm.

That evening he and his wife went off to dinner with the Swedish literary intelligentsia while I went with my fellow laureates in science to sup at the elegant naval officers' mess room on Stockholm Harbor at Skeppsholmen.

The next morning I got a sneak preview of the grandeur of the concert hall as my fellow laureates and I rehearsed the choreography of receiving a prize from the king's hands later that evening. As it usually is for most, this was to be my first experience of white-tie formality, and I was a bit self-conscious about how I looked. Betty, Dad, and I left the hotel at 3:45 P.M. to have more than enough time for me to join the backstage lineup. Precisely at 4:30 P.M. fanfare announced the arrival of the king and queen, who entered with their royal entourage and walked to their front-of-the-stage seats as the Stockholm Philharmonic Orchestra played the royal hymn. Then, with the trumpets again blaring, Max, John, Francis, Maurice, John Steinbeck, and I entered and took our seats near the front of the stage.

Before the king awarded each of the prizes, appropriate academicians read descriptions in Swedish of our respective accomplishments. To let us know what was being said, translations of their speeches had earlier been given to us. As the king handed each of us our leatherbound, individually decorated citations and gold medals, he also gave us checks in the amount of our individual shares of the prize money.

From the concert hall, we went directly to Stockholm's massive 1930s city hall for the Nobel banquet, which was held in the Golden Hall. Running the entire length of the beautiful room with vaulted ceilings was a very long table where all the laureates were seated with their spouses as well as the royal entourage and members of the diplomatic corps. Placed at its center facing each other were the king and queen. I was seated on the queen's side. While Max, John, Francis, and John Steinbeck all had princesses next to them, my conversation bits were to be alternately directed to the wives of Maurice and John Steinbeck. Talking across the table made no sense, because of both its width and the alcohol-enhanced din created by more than eight-hundred celebrants. During the dinner, the chairman of the Nobel Foundation, Arne Tiselius, proposed a toast to the king and queen; the king, in turn, proposed a minute of silence to honor Alfred Nobel's grand donation and philanthropy.

As soon as dessert was finished, John Steinbeck went to the grand podium overlooking the hall to deliver his Nobel address. In it he emphasized man's capacity for greatness of heart and spirit in the endless war against weakness and despair. The Cold War and the existence of nuclear weapons silently lurked behind his message of the writer confronting the human dilemma. He saw humans taking over divine prerogatives: "Having taken god-like powers, we must seek in ourselves for the responsibility and the wisdom we once prayed some deity might have." Ending his oration, he paraphrased St. John the Evangelist: "In the end is the Word, and the Word is man, and the Word is with men."

I became increasingly nervous and could not listen attentively, since in just a few minutes I was to be up on the podium to offer the response of the laureates in physiology or medicine. I hoped my extemporizing would rise above platitudes. Only after I was back at my seat did I relax, knowing that I had spoken from the heart. I was pleased at my last sentences, in which I had aimed for the cadence of one of JFK's better speeches. Graciously Francis then passed across the table his place card with a note on the back: "Much better than I could have done.—F." I could then enjoy John Kendrew expressing his joy at being part of a group of five men who had worked and talked together for the past fifteen years and could now come together to Stockholm on the same happy occasion. Then the party moved to the floor below for dancing, most of it done by the white ties and gowns of the Karolinska medical students.

Late the next morning the laureates in science gave their formal Nobel addresses. Francis, Maurice, and I were allotted thirty minutes each. It was not an occasion for questions from our audience of mostly fellow scientists. At seven-thirty that evening, I went alone to the palace for a second royal reception at which protocol had again somehow failed to place me beside a princess. This time I was between the wife of the Swedish prime minister and Sibylla, the wife of the Crown Prince Gustav Adolf, who had died in a 1947 airplane crash when his

M. WATSON:

Your Majesties, Your Royal Highnesses, Your Excellencies, Ladies and Gentlemen.

Francis Crick and Maurice Wilkins have asked me to reply for all three of us. But as it is difficult to convey the personal feeling of others, I must speak for myself. This evening is certainly the second most wonderful moment in my life. The first was our discovery of the structure of DNA. At that time we knew that a new world had been opened and that an old world which seemed rather mystical was gone. Our discovery was done using the methods of physics and chemistry to understand biology. I am a biologist while my friends Maurice and Francis are physicists. I am very much the junior one and my contribution to this work could have only happened with the help of Maurice and Francis. At that time some biologists were not very sympathetic with us because we wanted to solve a biological truth by physical means. But fortunately some physicists thought that through using the techniques of physics and chemistry a real contribution to biology could be made. The wisdom of these men in encouraging us was tremendously important in our success. Professor Bragg, our director at the Cavendish and Professor Niels Bohr often expressed their belief that physics would be a help in biology. The fact that these great men believed in this approach made it much easier for us to go forward. The last thing I would like to say is that good science as a way of life is sometimes difficult. It often is hard to have confidence that you really know where the future lies. We must thus believe strongly in our ideas, often to point where they may seem tiresome and bothersome and even arrogant to our colleagues. I knew many people, at least when I was young, who thought I was quite unbearable. Some also thought Maurice was very strange, and others, including myself, thought that Francis was at times difficult. Fortunately we were working among wise and tolerant people who understood the spirit of scientific discovery and the conditions necessary for its generation. I feel that it is very important, especially for us so singularly honored, to remember that science does not stand by itself, but is the creation of very human people. We must continue to work in the humane spirit in which we were fortunate to grow up. If so, we shall help insure that our science continues and that our civilization will prevail. Thank you very much for this very deep honor.

The transcript of my Nobel speech

daughters, the princesses, were still young girls. I found it easier to converse with the prime minister's wife than with Sibylla, whose native language was German. Sibylla ate practically nothing, perhaps imagining her still comely figure metamorphosing into the stereotypical one for royal consorts of the past century.

Before lunch the next day at the American ambassador's residence, I was taken to the Wallenberg's family Enskilda Bank to exchange my advance check of 85,739 kroner for one denominated in dollars, approximately $16,500. Earlier at Nobel House, I had been given a bronze copy of my gold Nobel Medal that I could safely leave lying about my desk. There had been past thefts of the gold originals, and I was urged to keep it in a bank vault. Seemingly hundreds of photos from the past days' festivities were then shown so that I could order copies of the ones I wished. Immediately my eye alighted on one of Francis and Princess Désirée, sitting across from me at the Nobel banquet.

Ambassador J. Graham Parsons greeted me graciously, giving no sign of the hawkish inclinations that had reputedly caused his recent banishment from the Washington corridors of Southeast Asia decision making. Also welcoming us was our embassy's number two man, Thomas Enders, whom I asked if he was related to John Enders, the Harvard Medical School's polio specialist who had won the Nobel eight years earlier. In fact, this Enders was the Nobel laureate's nephew, happily no longer living behind the Iron Curtain as a junior diplomat in Poland.

Nobel Week concluded traditionally on Saint Lucia's Day. Like all the laureates, I was awakened by a girl in a white robe and a crown of flaming candles, singing the Neapolitan hymn that long ago became virtually synonymous with this Swedish winter festival. With our father departing that afternoon for a week in France, Betty and I again put on formal finery for the Luciaball of the Medicinska Föreningen. At dinner reindeer was served as the main course. Afterward our party moved on to a much smaller private affair that let me banter long with Ellen Huldt, a pretty dark-haired medical student, with whom I then arranged to have dinner the next night.

Before getting into a taxi to fetch Ellen, I penned a letter to President Pusey, telling him of my visit that afternoon to the Royal Palace to see Princess Christina. With my diplomatic escort, Kai Falkman, I entered one of its private reception rooms to find her with her mother, Sibylla. Over tea and cakes, I related how much I enjoyed teaching the lively students of Harvard and Radcliffe and assured the mother that her daughter would greatly enjoy a year at Radcliffe. After returning home I sent back to Sweden several copies of Harvard's newspaper, the *Crimson,* to let Christina have a feel for Harvard undergraduate life.

As Nobel week ended, I was to depart for a visit to West Berlin arranged by the State Department, where my lecture before its scientists was to be yet another reaffirmation of the United States' unswerving commitment to those peoples trapped by the Cold War. Before flying there via Hamburg, I spent my last night in Sweden with John Steinbeck and his wife at the studio home of their friend the artist Bo Beskow. Liking *Ballet School,* one of his semifigurative blue paintings, I found its price to be within my somewhat improved means and arranged for it to be sent to Harvard. It long hung on the wall of the Biological Labs library.

In Berlin, I stayed for three nights at the guesthouse of the Freie Universität, a postwar creation sited among the buildings that once housed many of Germany's best scientists before Hitler. Until the Nazis came to power, Leo Szilard and Erwin Schrödinger had lectured on quantum theory there, with the voice of Einstein always figuring prominently in any discussion. Now of the past giants, only the Nobel Prize–winning biochemist Otto Warburg remained.

Soon to greet me was Kaky Gilbert, who the year before her Radcliffe graduation had assisted Alfred Tissières and me with experiments on messenger RNA. As a student at the Freie Universität, she would be showing me West Berlin. Unfortunately, she could not get permission to join me on my half day's visit to East Berlin, where I was surprised by the extraordinary Hellenic and Assyrian collections of the vast Pergamon Museum. Kaky did, however, accompany me to lunch at the residence of the head of the American mission in West Berlin. There I met the Prussian-acting Otto Warburg, whose legendary contribution to enzymology made him the most talked-about biochemist of our time. Though half Jewish, Warburg's longtime interest in cancer had led Hitler, always paranoid about contracting it, to let him continue working in Berlin throughout the war. He told me that my Harvard colleague George Wald was much too interested in philosophy, in contrast to his total lack of interest in it. Later, when

Kaky and I dined at what proved an all too typical German restaurant, I was again in throat agony. So we did not stay out long, going back to Dahlem on the underground train running out to West Berlin's southwestern suburbs.

My next stop was Cologne, where Max and Manny Delbrück were spending the year helping its university establish an antiauthoritarian, American-style department in genetics. Though I could barely whisper, my throat again unbearably sore, Max nevertheless insisted upon my giving a scheduled speech, to the distress of many in the audience who surmised my discomfort. Fortunately, by the time I reached Geneva, my voice had returned, allowing me to go out to CERN, the big physics laboratory then led by Max's friend from Copenhagen days Vicky Weisskopf. Much of our talk was about Leo Szilard, who had just flown back from Geneva to New York. Szilard wanted Vicky and John to set up a CERN-like multinationally funded European molecular biology lab after the model of Cold Spring Harbor Laboratory. Ideally Leo wanted the new lab to be in Geneva, but he would accept one on the Riviera. It would create an alternative intellectual home for him were the United States to tack right politically in response to even greater Soviet military threats.

Alfred and Virginia Tissières had by then arrived from Paris to spend the holidays with their family in Lausanne. I went up to ski with them just before Christmas at Verbier, the new ski resort in Valais that Alfred's brother Rudolph had helped develop. Soon to arrive were my new friends from Stockholm, Helen Friberg and Kai Falkman, who planned to be there through the New Year. By then, however, I would be with the Mitchison family at their home in Scotland. Though London was covered by snow, green grass still surrounded the airport near Campbeltown where my small British European Airways puddle jumper landed on New Year's Eve morning. From there I was driven some twenty miles to their remote Carradale home, arriving very weary from the journey. After three days of brisk highland walks, I flew back to the States via Iceland.

Back at Harvard I found on my desk a letter from President Pusey acknowledging my letter about Princess Christina, which he had passed on to Radcliffe's president, Mary Bunting. Appropriate application forms were dispatched to Sweden, and Christina promptly filled them out and asked her school, L'École Française, to send her academic records. News that she might be coming to Radcliffe first broke in the *Boston Globe* in mid-March, with an official announcement coming from the Royal Palace in early April. A day later, the *Crimson* asked about my role in her admission, and my attempt at humor badly backfired. To my embarrassment, the next day I read the words, "I didn't encourage her to come any more than I would encourage any pretty girl." I could only hope that that day's edition never got to Stockholm. In any case, I had further reason to believe I had made the right choice in moving to Harvard. Princesses don't go to Caltech.

Remembered Lessons

1. Buy, don't rent, a suit of tails

Though you may believe you will have no further occasion to wear a suit of clothes befitting an orchestra conductor, winning a Nobel Prize is too singular an occasion for a hired suit of clothes. Furthermore, if your career stays at a high level, you may be invited to a subsequent Nobel week when one of your protégés wins. And keeping the outfit in your closet long after the festivities are over will serve to remind you what shape you once were.

2. Don't sign petitions that want your celebrity

The moment your prize is announced, you are seen as fair game for petitioners of worthy causes in need of well-known signatories. In lending your name to such appeals, you often find yourself outside your expertise and expressing an opinion no more meaningful than, say, that of the average accountant. You trivialize your Nobel Prize and make future uses of your name less effective. Much better is to do real good as opposed to symbolic good.

3. Make the most of the year following announcement of your prize

You have a lifetime ahead of you for being a past prize winner but only a yearlong window during which you are the celebrated scientist of the moment. While everybody respects Nobel laureates, this year's winner is always the most sought-after dinner guest. In Stockholm this year's honoree is treated like a movie star by the general population, who will ask even an otherwise obscure chemist for an autograph. As with the Miss America pageant, the announcement of the next winner will decisively mark the end of your reign as this year's science star.

4. Don't anticipate a flirtatious Santa Lucia girl

Much fuss is made after your arrival for Nobel Week about the pretty girl who will wake you up on Santa Lucia Day and sing the traditional song. Alas, she will not be alone, and very possibly she will be accompanied by one or more photographers expecting you to smile as you hear the Neopolitan tune that only sun-deprived Swedes could mistake for a carol. The moment her singing stops, she will be off to another laureate's room, leaving you several hours more of darkness to endure before the winter sun peeks above the horizon.

5. Expect to put on weight after Stockholm

Masses of invitations will come to you during your inevitable bout of post-Stockholm withdrawal syndrome. You may find yourself banqueting as a second profession, accepting invitations to places it never would have occurred to you to go before. I still remember well an excellent dinner in Houston at its once classy Doctors' Club, before the Texas oil capital put itself on the map of high-powered biomedical research. I remember glaring foolishly at a giant ice sculpture on the table, knowing it would not long honor my existence. When your hosts embarrassingly overstate your importance, it's easier to accept second helpings than to keep up conversation.

6. Avoid gatherings of more than two Nobel Prize winners

All too often some well-intentioned person gathers together Nobel laureates to enhance an event promoting his or her university or city. The host does so convinced that these special guests will exude genius and incandescent or at least brilliantly eccentric personalities. The fact is that many years pass between the awarding of a prize and the work it acknowledges, so even recently awarded Nobelists have likely seen better days. The honorarium, no matter how hefty, will not compensate you for the realization that you probably look and act as old and tired as the other laureates, whose conversation is boring you perhaps as much as yours is boring them. The best way to remain lively is to restrict your professional contact to young, not yet famous colleagues. Though they likely will beat you at tennis, they will also keep your brain moving.

7. Spend your prize money on a home

A flashy car that costs more than it's worth is bound to give even your best friends reason to believe demi-celebrity has gone to your head and corrupted your values. Show them that the somewhat richer are not so different and you are still one of them. A bigger home will only put you in the same league as your university president, whom no one can reasonably envy.

11. MANNERS DEMANDED BY ACADEMIC INEPTITUDE

FROM the moment of my Nobel Prize, I took comfort in expecting a larger than ordinary annual salary raise. Over the past two years, I had twice received an annual increase of $1,000, so when I opened the small envelope coming on July 2 from University Hall I expected to see a $2,000 increase. Instead, the historian Franklin Ford, Bundy's successor as dean of the Faculty of Arts and Sciences, informed me that for my first time at Harvard I was to receive no raise at all. Instantly I went ballistic and let all my immediate friends know my outrage. Was an administrative blunder to blame for Harvard's failure to acknowledge the windfall of prestige that I had provided or did President Pusey want to send a message that celebrities had no place on his faculty and should consider going elsewhere?

Venting my wrath to student friends who wrote for the *Crimson* would be fun but likely to backfire and generate the official reply that Harvard never could adequately reward all the important ways the faculty enrich the academic milieu. Instead I talked to Harvard's best chemist, Bob Woodward, who was himself bound to receive the Nobel Prize soon. Attempting to calm me down, he told me he thought Harvard's failure to reward me reflected bad judgment on the part of our mediocre president as opposed to a deliberate insult. Bob offered to write Franklin Ford that if he were similarly treated, he would feel equally upset toward those who led Harvard. Later, Franklin Ford called me to his office to say that no insult had been intended—rather, priority had been given to rewarding other professors whose salaries

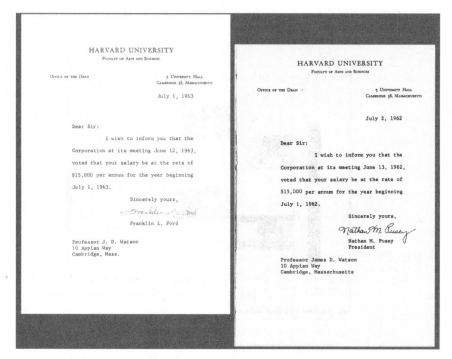

The letters that bookended the year I got a Nobel Prize but no raise from Harvard

were particularly low. The following year my salary went up by $2,000. My Spartan existence at 10 Appian Way, then as before, had allowed me routinely to spend less than I was earning. I thought about money only when I wanted to acquire for the walls of my apartment a painting or drawing beyond my means. Still, I would have been $1,000 poorer before taxes every subsequent year had I not spoken up about my displeasure.

More crucial to my morale than salary was how the science in my lab was going. Here I had cause for pleasure in the quality of my latest batch of graduate students—John Richardson, Ray Gesteland, Mario Capecchi, and Gary Gussin. With messenger RNA discovered, they knew how to proceed on their own. Underlying many of their successes was increasing use of phage RNA chains as templates for protein synthesis. To start us off, Ray Gesteland worked with Helga Doty to determine the molecular character of the RNA phage R17, whose RNA component of only some three thousand molecules most likely coded for only three to five different protein products.

The key surprise of the summer of 1963 was finding that RNA phages start their multiplication cycle through attachment to sex-specific thin filaments (or pili) coming off the surfaces of male *E. coli* bacteria. Such filaments are absent from female *E. coli* cells, explaining the until then mysterious fact that RNA phages grow only on male bacteria. Doing the electromicroscopy was Elizabeth Crawford, a summer visitor from Glasgow, with her molecular virologist husband, Lionel. Soon after their arrival, the three of us went up to the White Mountains, where we unintentionally aroused the ire of a nest-guarding goshawk that repeatedly dive-bombed us as we came down from a not undemanding walk up to the four-thousand-foot Carter's Dome.

Just before Labor Day, I flew to Geneva on my way to lecture at a NATO-sponsored molecular biology summer school at Ravello, Italy, across the bay from Naples. Among the other lecturers were to be Paul Doty, Fritz Lipmann, Jacques Monod, and Max Perutz. My high-ceilinged room in the Villa Cimbrone would have been perfect but for the nightly ravages of mosquitoes. Good fortune gave me as one of the sixty students the young German protein chemist Klaus Weber, then experimenting on the enzyme β-galactosidase for his Ph.D. at Freiburg. Klaus had come to Ravello to broaden his knowledge of nucleic acids and by the two-week program's end he'd accepted my invitation to come the following year to work on RNA phages in my Harvard lab.

At completion of the summer session, Leo Szilard flew down from Geneva to help lead further discussions about establishing in Europe a meeting and course site similar to Cold Spring Harbor Laboratory in New York. In Europe primarily to promote his latest scheme for preventing the nuclear annihilation of the planet, Leo came to Ravello on his way to a Pugwash disarmament meeting in Dubrovnik. Among those also briefly staying at the Villa Cimbrone were Ole Maaløe from Denmark, Sydney Brenner and John Kendrew from Cambridge, Ephraim Katchalski from Israel, and Jeffries Wyman, now living in Rome. By the end of the two-day meeting, widespread support existed for forming a European Laboratory of Fundamental Biology as well as a European Molecular Biology Organization of some one hundred to two hundred leading European biologists.

In a Rome art gallery on my way back, I lacked the courage to buy an almost affordable surrealist painting by an artist then unknown to me, Victor Brauner. During the following days, however, I acquired a small Paul Klee drawing from Gallery Moos, located below Jean Weigle's Geneva flat, and several André Derain drawings from Galerie Maeght in Paris. A week later, Princess Christina saw my new works of art at a small Sunday afternoon party I gave to help introduce her to Harvard life. Accompanying her was Antonia Johnson, the daughter of a leading Swedish industrial and shipping family, also to spend the coming year in the Radcliffe quad. To make the occasion more friendly, I invited some of my students, particularly those doing undergraduate research. But forty-five minutes later, when Christina and Antonia went off to another welcoming occasion, I remained unsure whether we would have reason to greet each other with more than a nod in passing during the year ahead. Neither Christina's nor Antonia's course choices were likely to bring them to a Harvard science building. On the other hand, Christina was likely to be friendly with my Radcliffe friend and future tutee Nancy Haven Doe, who had gone to the Spence School. She came from the New York social scene, which Radcliffe's first princess was bound to sample.

In the meanwhile, I became increasingly immersed in Biology Department politics. Carroll Williams was no longer chairman, having been succeeded by the into-Harvard-born behavioral biologist Don Griffin, who specialized in bat navigation. After being a Harvard junior fellow, he quickly rose in the academic ranks at Cornell before being called back in 1956. Don had natural affinities with Harvard's organismal biologists, so it was not anticipated that his first major goal as chairman would be to reduce the power of Harvard's separately funded biology museums, such as the Museum of Comparative Zoology and the Harvard Herbarium. Until this landmark turnabout, tenured museum scientists not only chose future museum curators but also had a say as to appointments to the Faculty of Arts and Sciences (FAS).

Back in June 1963, the tenured faculty, following Don's lead, voted that only professors supported by FAS monies would have automatic voting rights. At the same time, they opened up the possibility of key museum members having three-year terms on the Committee of Permanent Professors if so approved by two-thirds of the FAS-funded faculty. In this way, widely respected museum scientists such as the evolutionary biologist Ernst Mayr would retain voting rights. Carroll Williams, who still held much political power in the department, later tried hard to have the decision reversed. If the museum deadbeats were removed from the department scene, Carroll would no longer be seen as a neutral party in the tension between old-fashioned organismal biologists and the new group of molecular biologists. Instead he would be perceived as the leader of a conservative biology caucus openly determined to stop DNA-centered work from ever dominating Harvard's biology. It remained unclear whether Carroll would prevail until the fall of 1963, when a letter came from Franklin Ford reaffirming the opinion that only faculty paid by the Faculty of Arts and Sciences should vote on their respective department appointments.

These sensible voting qualifications, however, did not adequately guarantee first-rate appointments to the biology faculty. At least one-third of the Biological Laboratories remained unchanged since its construction thirty years earlier. Particularly out of date were its teaching labs, library, animal quarters, and machine shops. Moreover, the practice of relying on the senior faculty's government grants to renovate the out-of-date labs of incoming junior faculty put the latter cohort in a servile role. New funding would have to materialize from the Faculty of Arts and Science, not just from federal grants, if Harvard was to hold its own with Stanford, Caltech, MIT, and Rockefeller. Keith Porter, Paul Levine, and I prepared a report on facilities, which Don Griffin passed on to Franklin Ford. In it we outlined three possible scenarios for action. The first proposed extensive remodeling of the Biolabs; the second, a new five-story wing to the east; and the third, a ten-story building whose site demanded demolition of the historic brick Divinity School residence hall on Divinity Avenue, to the front of the Biological Laboratories.

A message soon came back from University Hall that no monies existed for construction of new Biology Department facilities. Any expansion of the biology faculty would have to occur within the pre-existing confines of the Biological Laboratories. This rebuff had an unexpected positive consequence. Administrative approval would be fastest for scientists already in situ, whose space requirements could be met by relatively inexpensive renovations of existing labs. Thus promoting Wally Gilbert to tenure was to prove much easier than we'd guessed the year before.

Wally still had a heavy physics teaching load and was supervising the Ph.D. theses of several graduate students. Only after the August 1961 Moscow Biochemistry Congress did Wally use most of his free time for experimentation in molecular biology. He first demonstrated that single poly U molecules serve as templates for several ribosomes simultaneously. Then he revealed the presence of transfer RNA molecules at the carboxyl ends of growing polypeptide chains. Currently he was showing that the attachment of streptomycin to ribosomes causes misreading of the genetic code. Even with these demonstrations of extraordinary talents as an experimentalist, I feared great difficulty getting him appointed to the biology faculty. In their eyes, Wally and I were too similar in our semiobjectionable objectives.

Luckily, Paul Doty soon orchestrated an arrangement to give Wally tenure, not as a member of the Biology Department but as a member of the Committee on Higher Degrees in Biophysics. Toward this end I wrote to Arthur Solomon, the long-reigning head of biophysics at Harvard, that Wally was in the same league, intellectually and experimentally, as Seymour Benzer or Sydney Brenner. Franklin Ford soon guaranteed funds to back the new tenured biophysics slot, and Wally's promotion breezed through the ad hoc committee early in April 1964.

The approval process for Wally occurred when I was in Paris to lecture at an April meeting marking the fiftieth anniversary of the French Biochemical Society. There, for the first time, I announced evidence for the existence of two ribosomal sites that specifically bind transfer RNA. One site I called A in view of its function to bind incoming amino acid transfer molecules in the presence of messenger RNA codons. The second site I called P since it holds the growing polypeptide chains to the ribosomes. In ways yet to be determined, growing polypeptide chains immediately moved from A sites to P sites after each new round of peptide bond formation. Toward the end of my lecture, I emphasized that we also expected to find along mRNA molecules signals for starting and stopping polypeptide synthesis. Our failure so far to find them may have reflected the fact that current cell-free systems for protein synthesis used only synthetic RNAs as templates. These contain no signals. Synthetic molecules likely worked as good templates only through mistakes in codon reading. Fearing afterward that my talk would be perceived as speculative, I was hugely relieved when François Jacob, never one for idle praise, called my lecture one of the best he had ever heard.

I was at the time very proud to be a senior fellow of Harvard's Society of Fellows. The main task of the eight senior fellows each year was to choose a similar number of junior fellows whose terms ran three years. We were also expected to dine each Monday with the junior fellows in the society's wood-paneled quarters in Eliot House. The society's reputation for exemplary minds was still deserved. The evolutionary biologist Jared Diamond and Nobel Prize–winning chemist Roald Hoffman were both selected in 1962.

It was much more fun, I found, to sit next to junior fellows rather than senior fellows; utterly painful was getting caught beside either the acerbic but shy critic Harry Levin or the overly polite classicist Herbert Bloch. And though the philosopher Willard V. O. Quine may have been the brightest of all, his voice came alive only when talking about maps. One such evening per year would have been enough for me. Dinners became more agreeably animated after the economist Wassily Leontief took over as chairman in July 1964 and brought about the appointment as senior fellow of Boston's senior judge, the gossipy Charles Wyzanski, whose interests—intellectual as well as social—went far beyond the Harvard scene.

The historian Crane Brinton still presided over the society when I suggested inviting Princess Christina to one of our Monday night dinners. Soon I got the message that her royal presence might distract from its intellectual purposes. We had recently run into each other at the usually jammed coffee shop across from Widener Library. Antonia Johnson was with her, so the three of us shared a booth for lunch.

Much of our conversation centered on Antonia's current boyfriend in Philadelphia, who was doing biophysics research at the University of Pennsylvania. Several weeks later, I met Christina again at a waltz evening at the Parker House Hotel, where she discovered that I did not belong on a dance floor. I was the unlikely escort of the full-bodied blond bombshell Sheldon Ogilvy, whose college education in New York City had recently gone astray. Somehow she was on the guest list for this minor monthly Boston society dance. Exuding the happy insouciance of a Truman Capote heroine, Sheldon sipped Brandy Alexanders when we sat together at the Club Casablanca beneath the Brattle Theatre.

Early in May, Christina and I were both at a Saturday night dance in Locust Valley, on Long Island's Gold Coast. It marked the twenty-first birthday of Deming Pratt, Nancy Doe's Radcliffe roommate. A marriage bond once connected Deming's branch of the oil-rich Pratt family to the royal Bernadottes of Sweden. Nancy told me I was to come, so I arranged to stay that night at the Cold Spring Harbor Laboratory, only a fifteen-minute drive to the east. The scotch and soda I consumed over a pre-dance supper at the Piping Rock Club gave me the courage to once again subject Christina to my lack of rhythm. Much of the evening I gossiped with the Texas department store heiress Wendy Marcus and her escort, a *New York Times* Latin America correspondent. Leaving after most guests had already departed, I absentmindedly drove off without turning on the headlights. Immediately I was pulled over by a local police officer, who mercifully only told me to stay on the shoulder until my head cleared. The times then were much more foolishly forgiving than today of intoxication behind the wheel. Later I shuddered at the thought of the publicity that would have ensued had the policeman been more hard-nosed and done his duty to haul me in.

The arrival of summer left Sheldon Ogilvy no reason to remain in Cambridge and she became ensconced at 336 Riverside Drive in New York City. Though no relation of the advertising whiz David Ogilvy, she got hired as the receptionist at Ogilvy and Mather's offices on Madison Avenue. I popped in to see her when I came down late in July for the 1964 International Biochemical Congress at the Hilton Hotel. Francis Crick was there to give one of the congress's keynote speeches. In my talk, I slipped in a slide showing a photo from the Nobel Prize dinner of Francis apparently peering inappropriately at Princess Désirée. Later, to make amends, I introduced Francis to Sheldon at a lighthearted dinner at the piano bar restaurant on the roof of the St. Regis. The next day Francis telephoned her at work to get her address and phone number for his next trip to the States. I only learned this in early December, when Sheldon wrote telling me that she would be having lunch with Francis again on his way back to England, and that she first felt a bit odd in accepting his attention and then a bit odd about being so scrupulous, given that the three of us had had so much fun together at the St. Regis.

No longer a receptionist, Sheldon was on to teaching foxtrots to what she called "little WASP girls" in Westchester so they would look presentable when they suddenly blossomed into debutantes. In her note, she expressed regret at backing out with virtually no notice from an October weekend with me at Cold Spring Harbor. The previous night's activities, she explained, had left her too black and blue to appear in public. Discretion kept me from asking for details. Less athletically, she was immersed in Virginia Woolf and wanted my help to win readmission to Barnard and dispel the cloud hanging over her since her abrupt withdrawal. Although I could see that her cause would benefit from a supportive letter, it was not all clear to me what I could credibly write.

Over the fall of 1964, a growing faculty consensus, strongly encouraged by Franklin Ford, emerged for placing the long-running under-graduate concentration in biochemical sciences under the jurisdiction of the Committee for Higher Degrees in Biochemistry. Most biochemical science majors historically aimed for medical school, and many members of its Board of Tutors, correspondingly, had medical school affiliations. In 1958, the microbiologist Alvin Pappenheimer came to Harvard from New York University Medical School to become the board's head tutor, replacing the veteran John Edsall. Now Pap wanted to resign since he had just been appointed master of Dunster House.

The proposed merger would allow the junior faculty members to rotate in and out of the head tutor position without the dean's having to create a tenure slot for each new appointee.

To win support from those who wanted a separate biochemistry department created immediately, Franklin Ford gave permission to start a search for a senior biochemist or molecular biologist, a boon to these disciplines. Equally important, he and President Pusey promised a new science building to be sited between the Converse Memorial Lab and the Biological Laboratories. The official formation of the new Committee on Biochemistry and Molecular Biology (BMB) was announced at the February 1965 meeting of the Faculty of Arts and Sciences. Its new chairman was to be John Edsall, signifying to all a seriousness of commitment.

Widespread enthusiasm then existed for bringing the Canadian-trained M.D. David Hubel, who was associate professor of neurophysiology at Harvard's Medical School, to the Biology Department with tenure. Current experiments by Hubel and his Swedish-born collaborator Torsten Wiesel were radically advancing knowledge of how the visual cortex is organized. To allow their collaboration to continue, Don Griffin proposed giving Wiesel an appointment as a senior research associate in biology. Especially impressed by their accomplishments was Francis Crick, who now used his role as a nonresident fellow of the Salk Institute to meet frequently with Hubel and Wiesel in La Jolla. Attracting Hubel and Wiesel would be a massive step forward for the Biolabs, stamping it indelibly as a place of high-level biology. Speculation already existed that they were bound to receive a joint Nobel Prize. Common sense thus dictated that Wiesel also be offered a tenured position. But we were told that the dean could not now create a new tenured slot for him, especially since he was said to have no interest in teaching undergraduates.

I soon had an opportunity to become better acquainted with Hubel and Wiesel through a late winter visit to the Salk Institute in La Jolla. The occasion for coming to the San Diego region was a three-day gathering on the role of genes in the immune response, held at Warner Hot Springs near Palomar Mountain and its big telescope. There Norbert Hilschmann from Lyman Craig's lab at Rockefeller University put up a slide showing amino acid sequences from antibody-like Bence-Jones proteins found in victims of multiple myeloma. He took care not to let this slide stay on the screen long enough for its data to be copied down by his rival at Rockefeller, Gerry Edelman, then also in the audience. Enraged by this act of bad form, Max Delbrück rose and denounced Hilschmann. But those of us who knew Gerry well could see Hilschmann was in a no-win situation.

In deciding at the last moment to extend my California visit for an additional week, I would be violating a long-standing Arts and Sciences rule that during term the president and fellows must approve all visits of more than a week away from Harvard. But I was not scheduled for lectures during the time in question, and asking at the last moment for approval might delay my departure to Warner Hot Springs. So I decided simply to tell Don Griffin that I was to be away for some two weeks. The thought never occurred to me that he would see the need to tell Franklin Ford. But this he did, and I only learned of it at the Salk Institute in the middle of watching Hubel and Wiesel in action. They were among the first to know of my instant rage when I got a phone call from Don telling me that President Pusey wanted me immediately to return to Harvard. I was being treated like an AWOL soldier. Deeply upset, I told Don that while Pusey might get satisfaction from humiliating me, he was giving Hubel reason to wonder why he should consider giving up Harvard Medical School, where he could travel as he saw fit, to move to the Faculty of Arts and Sciences and be governed by rules more befitting teenagers than serious scientists. Don phoned the next morning telling me I could stay. Until then, only the compound curse "F—— Harvard and f—— Pusey" went through my brain, and I vocalized it openly many times as the day turned to night.

Upon my return to Harvard, Paul Doty gave me the dope that Kenneth Galbraith had also had a run-in with Pusey over traveling during term. Denied a request to spend a winter without teaching commitments skiing in Gstaad, Ken obtained a doctor's note that the break was medically necessary to prevent undue stress on his circulatory system. Not wanting to risk a fight that Ken might make public, Pusey caved in. The Harvard Corporation was still embarrassed about almost having blackballed Ken after World War II for supposed left-wing economics. I feared the line Pusey had taken with me was his way of reasserting authority over his faculty. Needless to say, I was apprehensive on receiving Franklin Ford's July 1 note concerning my salary for the coming year. To my relief, I got a $1,000 increase.

Three months earlier, I had been even more gratified to read another letter from Ford stating that Mr. Pusey and the Corporation wanted to reconsider aspects of policies concerning travel and salaries. They wanted the matter studied over the next year by a faculty committee whose work would be highly confidential. Would I consider chairing the committee, despite my plan to be away for much of the next year? I felt immensely vindicated and phoned Paul Doty. To my surprise, he did not seem happily surprised at the Corporation's turnabout. It was only several hours later, on my way to Paul's house, that I looked again at Ford's letter and suddenly noted it was dated April 1, 1965. Meeting his friend Dorothy Zinberg at the gate, I saw clearly that we were victims of the same hoax. I should have known that Harvard never would have changed course so dramatically, but I couldn't for the life of me work out how Paul had obtained the University Hall stationery.

At no time did my run-in with President Pusey make me want to leave Harvard. Nowhere else was I likely to get such a caliber of graduate students. Their latest triumphs involved besting two labs at Rockefeller University in understanding key features of protein synthesis. For more than a year, Norton Zinder's Rockefeller group and my RNA phage group had raced each other to find out how so-called nonsense bacterial suppressor strains misread mutant chain-terminating signals to generate biologically active polypeptide chains. By late spring Gary Gussin and Mario Capecchi wrote up for publication in *Science* that mutant tRNA molecules read "nonsense" signals as "sense" signals, thus winning the race.

Less than six months later, Capecchi and Jerry Adams showed that formyl methionine tRNA molecules initiate the synthesis of bacterial protein chains. Earlier I visited Rockefeller University to see whether Fritz Lipmann's big lab was following up the discovery of f-met-tRNA, made some months before in Denmark. Its existence might explain why so many bacterial proteins had methionine as their terminal amino acids. But Fritz was not thinking along these lines, and I left New York City knowing Jerry Adams would have no competition studying how protein synthesis starts. Soon Jerry discovered how to radioactively label f-met-tRNA molecules, allowing him and Mario to label the formyl groups at the ends of RNA phage proteins made in vitro. Their experimental results were sent off to the *Proceedings of the National Academy* just before I flew to London to spend December 1965 in Cambridge.

By then, virtually everybody in the Biological Laboratories knew that their best interests would be served if the BMB Committee rapidly converted into a genuine department. Uncertainty about being allocated space was causing qualms for prospective faculty recruits. Keith Porter had taken over as chairman after Don Griffin's three-year stint. Only months before the changeover Griffin had unexpectedly announced that he was resigning to move to Rockefeller University and its field station in Millbrook, some fifty miles north of New York City. In a panic, the Biology Department offered his tenured slot to Edwin Furshpan, who studied invertebrate synapses at Harvard Medical School in a lab nearby that of David Hubel and Torsten Wiesel. Behind this hurried offer were hopes that it would make David Hubel more inclined to move to Cambridge.

This ploy, however, didn't work. David, Torsten, and Ed eventually decided to remain at Harvard Medical School. The Pappenheimer microbiology tenure slot was also up for grabs, since Boris Magasanik had turned it down more than a year before to remain a member of the MIT biology department, whose future he saw no reason to question. Now the department was prepared to change tracks and offer the position to Renato Dulbecco, whose research on DNA tumor viruses was zooming forward. No one, however, was surprised when Renato declined it, knowing his research facilities in the soon-to-be-completed Salk Institute would be incomparably better than Harvard could offer in the Biolabs.

Wanting to pull off at least one coup of tenure acceptance, Keith

Porter enthused about recruiting the circadian rhythm expert Woody Hastings, from Illinois. Long a fixture of the Woods Hole summer scene, Woody was liked by all, and I went along in voting for his appointment though I saw his science as having little potential to make lasting ripples. As I was soon to leave the Biology Department to take on BMB stripes, I saw only ill will coming of opposing Woody on intellectual grounds. If the appointment was to be blocked, the move would have to occur at the ad hoc committee level. But the committee was composed of academics who thought biology teaching had to remain diverse, and the appointment went through.

At the January 11, 1966, meeting when the Biology Department recommended the Hastings appointment, it also made Wally Gilbert a member, opening the way for him to obtain space of his own that I assumed would be adjacent to mine. There was also much discussion about John Edsall's memorandum to Franklin Ford urging the speedy creation of a Department of Biochemistry and Molecular Biology. Keith said he would form a three-person committee to draft a statement that everyone anticipated would express the Biology Department's support for the split. A similar discussion occurred in the Chemistry Department, which likewise accepted the split, though it expressed regret they would lose from their ranks important figures such as Paul Doty. Believing the matter to be sufficiently important for outside analysis, Franklin Ford took John Edsall's specific proposals before a specially convened ad hoc committee that requested more precise details before ruling in favor of a subsequent revised proposal. Officially the Department of Biochemistry and Molecular Biology would come into existence on February 1, 1967, with its offices on Divinity Avenue in the large wooden house where the famed historian Arthur Schlesinger had been raised.

The impending creation of the BMB, however, did not put my mind at rest. While spending the spring in Alfred Tissières's lab in Geneva, I got wind that the biology faculty was proposing to move the geneticist Paul Levine onto the third floor, adjacent to my lab. Giving Paul this space would not only prevent Wally Gilbert from getting it but also limit the possibility of locating junior faculty members near Wally and me. As Dulbecco had just declined our offer, I wrote to Keith saying

In my office in 1967

that if new blood was to enter the Biolabs, it would have to come through the junior ranks.

The thought of their making Paul my neighbor, and effectively forcing Wally to move to the fourth floor, made my blood boil, and I told Porter so. Ever since Wally had gained tenure, our students had two mentors simultaneously—a unique research experience. Spirited conversations over coffee, lunch, and tea would occur much less frequently with our two groups on two different floors. Over the summer, talk of Levine's relocation stopped, leading me to hope that I had made enough of a stink to scotch it. But the plan was resurrected in September with Geoffrey Pollitt, the Biolabs' senior administrator, arguing that it would free up ten units of precious lab space. In my mind the same objective could be achieved by reducing Levine's domain, which now equaled mine in square footage, though with only half the personnel. Not until mid-December did Keith officially backtrack, offering Wally the same office and research space that in May he had proposed giving to Paul Levine.

Our new department was coming into existence not a moment too soon.

Remembered Lessons

1. Success should command a premium

Whether my Biology Department enemies helped orchestrate Harvard's failure to acknowledge my Nobel Prize with a respectful salary increase, I will never know. I had added visibly to Harvard's image capital in a way that by my lights merited more than the president's pro forma one-sentence congratulatory letter. Much bunk is peddled about money not being a prime motive of the academic. This does not, however, change the fact that salary is the means by which any employer expresses how much he values you; whether you need the cash or not, make sure your salary reflects your status.

2. Channel rage through intermediaries

Feelings of intense anger against university administrators are best conveyed through friends who share your feelings of mistreatment. Directly confronting your dean all too easily leads to words that burn bridges within your institution. It never makes sense to be seen as a hothead unable to see another person's point of view.

3. Be prepared to resign over inadequate space

Your colleagues won't know whether your raise is a thousand dollars more or less than theirs, but everyone can see what sort of work space you are assigned. It affects what you can accomplish and how you are perceived. Assigning equal space to all equally ranked academics sounds fair but leads to inefficiencies. Individuals at different stages of their career have different needs but giving or taking away space accordingly leads to complaints of favoritism, and so political rather than rational allocation is the norm. Yet if you find yourself denied the space you need to exploit a bright new idea or experimental breakthrough, you may be overtaken by competition elsewhere. Losing an important space request means either your talents are not recognized or your department is relatively indifferent to whether you stay or go. If you don't make a credible threat to resign, you will never know where you stand. Such moments are inherently stressful and never to be taken lightly. But in the Darwinian world of an academic department, if you don't create such crises, limited resources will surely go to gutsier colleagues.

4. Have friends close to those who rule

When Charlie Wyzanski learned over a Society of Fellows dinner that I had been summarily called back to Harvard while on a visit to California, he wrote to his friend at the Harvard Corporation, Thomas Lamont. In a private letter, Charlie expressed how asinine Harvard would have looked if the matter had leaked to Boston newspapers. And a word to the wise was sufficient.

5. Never offer tenure to practitioners of dying disciplines

In the 1960s, the Harvard Biology Department continued to make tenured appointments in fields such as development and plant biology, tired games not likely to rebound soon. Undergraduate teaching needs were invariably cited for such appointments. MIT, on the other hand, practiced attrition with these dying disciplines, leaving the teaching of them to untenured faculty. Consequently, by the mid-1970s, our academic rival began moving from behind to far ahead of Harvard in biology, with predictable effects on the quality of graduate students in both places.

6. Become the chairman

Most top university scientists disdain duties that take time from research. They see administration as a bore, and everyone wants someone else to be the department chairman. As a result of shirking responsibility, most science departments are less exciting places than they should be. The straw that stirs the drink counts for a lot. Dull chairmen make foolish choices when they assign the teaching of important courses and the use of precious department space and facilities. The wrong faculty members handle departmental seminars and keep the library buying journals that no one reads. Departmental meetings have no purpose droning on without addressing vital issues until there is no oxygen left in the room. Being chairman need not consume more than 10 percent of an intelligent professor's time, possibly less than he or she might waste griping about bad decisions made by others.

7. Ask the dean only for what he can give

Both saying and hearing no are unpleasant experiences, making the denier look ungenerous and the denied clueless or impotent. Everyone has a wish list, but when you ask the dean for something, make sure you have thought through how he could reasonably give it without taking a costly political hit. It is quite another matter, however, to ask the dean to petition the university for a change of policy you want. Although a negative reply is always annoying, here neither you nor the dean looks bad personally, and you will at least gain insight into the university's finances and priorities.

12. MANNERS BEHIND READABLE BOOKS

*T*he *Double Helix,* the story of how Francis and I found the structure of DNA, was published in February 1968—fifteen years after the discovery. That the tale's many unexpected twists should be revealed to the general public was long on my mind, but how to write them up did not crystallize until a spring 1962 dinner in New York City. There Francis and I were being honored with the Research Corporation Prize. Francis could not be there since he was in Seattle delivering three long-scheduled public lectures at the University of Washington. Abby Rockefeller invited me to spend the night at her parents' home, where I admired her father's large Derain fauve painting of the Thames and beheld her family's porcelain with less appreciation. I walked from East Sixty-fifth Street to the Ambassador Hotel on Park Avenue, then the venue of many such ceremonial affairs. On the dais I was next to Columbia University's literary polymath Jacques Barzun, known to me since my adolescence through his regular appearances on the CBS radio network.

Stimulated by Barzun's conversation, I used my after-dinner acceptance speech to tell the story of our discovery as a very human drama also featuring Maurice Wilkins, Rosalind Franklin, Erwin Chargaff, and Linus Pauling. My unexpected candor elicited much laughter and was later praised for allowing the audience to feel like insiders in one of science's big moments. Feeling jubilant as I walked back along Lexington Avenue to the Rockefeller home, I saw in my future the writing

Abby Rockefeller

With Cynthia Johnson in Radcliffe Yard

Wally Gilbert and I ride the rhino outside the Harvard Biolabs with Barbara Riddle and blue-eyed Pat Collinge.

of what Truman Capote would later call the "nonfiction novel." But since I was scheduled almost immediately to return to England to my mini-sabbatical at Churchill College, Cambridge, I could not see starting to write until my return to Harvard.

In London I initially hoped to use my half of the Research Corporation Prize to commission Francis Bacon to paint Francis Crick; I had recently seen one of Bacon's small portraits in Geneva, and it long lingered in my mind. But the Marlborough Gallery let me know that the Irish-born Bacon painted only close acquaintances. The fact that Francis and Odile had spotted the artist the previous summer in Tangier would not suffice. An hour later, I walked out onto Albemarle Street the contented possessor of one of nine copies of Henry Moore's bronze *Head of a Warrior*.

The first chapter of what later came to be called *The Double Helix* was written in Albert and Marta Szent-Györgyi's house on Cape Cod. The summer was coming to an end and my Radcliffe summer assistant, Pat Collinge, was about to join her Harvard boyfriend, Jake, at work on a novel farther out on the Cape. Since the start of summer, Pat's blue eyes and urchin dress made her seem the perfect muse to draw out of me the first pages of my DNA story without apparent effort. Happily, she agreed to be driven down to the Cape in my open MG TF to spend a morning as my typist at Woods Hole before going on to Wellfleet. But I had writer's block for several hours before the words came to me: "I have never seen Francis Crick in a modest mood." More than half the first chapter was typed out before her boyfriend showed up, leaving me wistfully contemplating how I would finish the rest of the chapter without the encouragement of her blue eyes. It wasn't so easy, and only once back at Harvard would I finish the final paragraphs, whereupon my secretary typed out a complete first chapter.

Events connected to my fall Nobel Prize conspired with the school year to keep me from writing any more until the following summer. It was then I spotted the engagingly pretty Radcliffe senior Cynthia Johnson, eating lunch under the big elm tree in front of the Biolabs. She was with our department's best-looking assistant professor, John Dowling, under whom she was doing summer research in George

Wald's vision lab. The next day I joined them for lunch and soon afterward was playing tennis with Cynthia on the Radcliffe courts. Too soon I learned that she had a steady boyfriend, Malcolm MacKay, who, having just graduated from Princeton, was in Europe before starting Harvard Law School in the fall. Until that time, however, we took day trips together, once to the beach at Nahant, where, scarily, I first noticed that ceremonial overeating had given me a slight belly, the first fat I'd ever observed in my midsection. Later, on Martha's Vineyard, I stayed several weekends at Cynthia's family home in Edgartown, learning not only that her artist mother had drawn the Duke and Duchess of Windsor but that her grandmother was a close friend of Emily Post, whose book on manners precipitated the decline of the WASP ascendancy in America. Thinking that my being a writer as well as a laureate might offer romance enough to dislodge Malcolm from Cynthia's heart, I spent the last two weeks of August writing two more chapters. But when the fall came and Cynthia brought Malcolm to my flat for my approval, I knew I could not compete with his sailboats.

By then my free moments were devoted to completing six short chapters on the replication of living molecules for a book to emerge in time for my forthcoming January lectures to talented high school students in Australia. George Gamow had lectured to this Sydney summer school the year before and highly recommended it as an excuse for being in the warm sun during January. I likewise looked forward to avoiding the East Coast doldrums over the Christmas/New Year holidays and so accepted the necessity of writing up simplified versions of my Biology 2 lectures.

One muggy August evening, on the plain wooden desk of my Appian Way flat, I wrote out, "It is very easy to consider man unique among living organisms. Great civilizations have developed and changed our world's environments in ways inconceivable for any other form of life. There has thus always been a tendency to think that something special differentiates man from everything else. These beliefs often find expression in man's religions that try to give an origin to our existence and in so doing to provide workable rules for conducting our lives. Just as every human life begins at a fixed time, it was natural to think that man did not always exist but that there was a moment of creation perhaps occurring at the same time for man and all other forms of life. These views, however, were first seriously questioned just over 100 years ago when Darwin and Wallace proposed their theories of evolution based on selection of the most fit."

The remainder of this introductory chapter came easily over the next several days, leaving me confident of getting the remaining five pieces done by the end of October. In that case, the book would be ready at the start of the Australian summer school. But pressing tasks for the President's Science Advisory Committee let me send off only three chapters ("Introduction," "A Chemist's View of the Living Cell," "The Concept of Template Surfaces"). Later John F. Kennedy's death in Dallas and my father's stroke soon after kept me from writing up the last three lectures. To my relief, my father's weakness on one side was not permanent, and by the time I took him down to Washington to be with my sister over the winter, he could walk to and from Harvard Square using a cane.

Despite a brief stopover in Fiji, I felt and looked haggard when I arrived in Sydney just after New Year's. The short brown beard that I had grown in September while lecturing in Ravello led a Sydney newspaper article to describe me as "haunted and Mephistophelean" and "as introverted as his host the physicist Harry Messel was extroverted." It went on to characterize me as difficult to entertain, a portrayal reflecting several dinner parties given for me by senior academics, not one graced by the face of a pretty girl. In desperation I suggested an evening of nightclubbing only to find that in prudish Sydney chorus girls still danced fully clothed.

My life improved dramatically when a reporter for the *Mirror* arranged for me to be led around the Paddington art scene by the young painter and critic Robert Hughes. In the Rudy Komen Gallery, I bought the large blue painting *Kings Cross Woman* by the former boxer Robert Dickerson and a de Kooning–like green-faced woman by the equally talented Jon Molvig. Then we popped into the Kellman Gallery, where I acquired a life-size wooden man from the Sepic River peoples of Papua New Guinea. Its painted Dubuffet-like face later sat across from me at my Biolabs desk. At last in high spirits, I looked forward to a second day of gallery hopping, but Hughes pulled out, making me suspect that women did not make him tick, a notion since copiously disproved.

Even before my arrival back at Harvard, I feared my Australian lectures and their written versions were for college, not high school, students and that audiences would get lost. My three completed chapters, I decided, would work better as the start of a little college-level text on how DNA provides the information that enables cellular existence. A chance meeting at the Wursthaus in Harvard Square led me to learn that MIT's molecular biologist, Cyrus Levinthal, was advising the new science textbook company W. A. Benjamin. The firm was eager to expand its initial physics and chemistry list to encompass biochemistry and molecular biology. Less than a week later, its main editor, the young Canadian Neil Patterson, came to my office to make me a Benjamin author, as he had earlier the über-physicist Murray Gell-Mann.

Soon after, I was in W. A. Benjamin's grubby New York City offices, above a bowling alley on Upper Broadway. My discomfort abated when I was told they would soon move to 2 Park Avenue, below Grand Central Station. Liking the way in which Neil Patterson sought out books by clever young scientists, I signed a contract that gave me a $1,000 advance for a 125-page book to be completed late in the year. In addition, I was offered options to buy five thousand shares of Benjamin stock; there seemed reason to hope the stock price would rise as new books rolled out. By then I had given my prospective book the title *The Molecular Biology of the Gene* (*MBG*). I was first tempted to call it *This Is Life,* a titular rejoinder to Erwin Schrödinger's *What Is Life?* On reflection, however, that would have been promising more than I could deliver.

In writing my new chapters, I used boldface sentences to summarize the main idea conveyed by paragraphs below it (e.g., "Molecules are restrictively sticky"; "Enzymes cannot be used to order amino acids in proteins"; "Template interactions are based on weak bonds

*In the right-hand corner of my office, the Sepic River wood carving
I bought shortly after winning the Nobel Prize supervises my labors.*

over short distances"). I hit upon "concept heads" as a teaching device when writing the chapter "A Chemist's Look at the Living Cell" before going off to Australia. They naturally emerged from lists of ideas I prepared in outlining what topics each chapter should include. Almost from the start I saw the need to expand upon my Australian chapters, coming up with snappy concept heads such as "The 25-year loneliness of the protein crystallographer."

Equally important to the final readability of *MBG* was the artwork done by the young Keith Roberts, about to begin his university studies. Early in 1964, Keith had come from England to work as a temporary lab technician prior to reading botany at Cambridge. When I happened to ask his opinion of my first draft chapters, he revealed that he had almost chosen to study art over science, and volunteered to do the necessary illustrations. As my manuscript steadily grew in length, Keith's preoccupation with drawing became full-time, and he continued to draw for me after commencing his freshman year.

At that time, Benjamin was using two-color printing and professional artists. Here I was lucky to have the New York painter Bill Prokus help me transfer Keith's artistic ideas into fixed artwork. Bill then had a studio on Twenty-third Street in Chelsea, where in addition to his own work he did commercial artwork for Benjamin. To speed Bill along, I began coming down regularly to New York City and staying at the Plaza, where the inside rooms never cost more than $20. Even so, Bob Worth, the steel heir who was Benjamin's financial officer, called on me to stop such visits, having read an unfavorable opinion of my then almost finished manuscript. Neil Patterson had sent it to the cell biologist Bob Allen at Dartmouth, who found it unsuitable for his students. Fortunately, Neil prevailed over Bob, and I did not have to stay at the dingy Chelsea Hotel, sited across from Bill Prokus's studio.

All my chapter drafts were greatly improved in editing done by the Radcliffe senior Dolly Garter. She had taken George Wald's first-year general-education biology course and so was exposed to the DNA way of thinking. That she was an English major interested in writing was a big plus, and she was able to change many a turgid phrase into freer-flowing language. My challenge became getting the chapters to the point that Dolly could understand them without recourse to Keith's illustrations, often not yet done. If Dolly could follow the words alone, I figured less bright students would not have difficulty with my arguments suitably illustrated. Dolly worked right through the fall of 1964 revising chapters as I wrote them. By then I had expanded *MBG*'s scope, adding the chapters "The Importance of Weak Chemical Interactions" and "Coupled Reactions and Group Transfers" for students coming from biology with weaker backgrounds in chemistry. Together with many later chapters, they needed constant rewriting to keep pace with the latest scientific advances. The countless revisions of the initial galley proofs led to most of *MBG* being completely reset. Even in page proofs I made many more changes than my publishers wanted, and they threatened to dock me with the charges involved. In the end, they never did so, realizing the value on balance of an up-to-the-minute book.

A wisp of pale, fragile flesh, the Brooklyn-born Dolly belonged to the circle of Harvard's literary magazine *The Advocate*. So I had her read the first chapters of my memoir about finding the double helix. The book's original title was *Honest Jim*, since Alfred Tissières earlier had reminded me that Maurice Wilkins's collaborator, Willy Seeds, had cynically addressed me as "Honest Jim" when in August 1955 he met Alfred and me by chance on a path in the Alps. Now, *Honest Jim* was my way to face head-on the controversial question prompting Seeds's cynicism, over whether Francis and I had improperly used confidential King's College data in working out the structure of DNA. I had in mind the way Joseph Conrad used *Lord Jim* to pose the basic question as to its hero's character.

Dolly's enthusiasm for *Honest Jim*'s first chapters encouraged me to get back to writing once the essential features of *MBG* were in place. By the time she and her boyfriend, a math major and fellow Brooklynite called Danny, graduated in June, Dolly had read half of *Honest Jim*. From her new publishing job with Van Nostrand in Princeton, she later wrote telling me that the Harvard *Advocate* would like to publish its opening chapters. Though the life of a scientist, in her opinion, was of necessity dull, the *Advocate*'s readership would benefit from reading about my exhilarating experiences.

At the time I had hopes that Houghton Mifflin would publish *Honest Jim*. The year before on a May evening I'd driven out to Beverly, to the elegant large square wooden house of Dorothy de Santillana, a senior editor at Houghton Mifflin, whose husband Giorgio's expertise was the history of science. We had met earlier through Dorothy's younger relative the Radcliffe graduate Ella Clark. Over dinner I enjoyed talking to the novelist Alberto Moravia's much younger wife, Dacia Maraini, who had just published a sexually charged novel of her own. In leaving, I gave several early chapters of *Honest Jim* to Dorothy. Soon she wrote me a flattering note saying that when my manuscript was more complete, I should show it to Houghton Mifflin.

As the spring 1965 term ended, I flew off to Germany to deliver three lectures. The first was in Munich, where I shared my first suckling pig in a large beer hall with the biochemist Feodor Lynen. Then his country's most accomplished biochemist, he went to the States at least twice a year to keep up as an enzymologist. Later that year, he would be receiving the Nobel Prize in Physiology or Medicine jointly with my Harvard colleague Konrad Bloch, for discovering how cholesterol is made in cells. I next traveled north, first to Würzburg and then to Göttingen, where the chemist Manfred Eigen's fast-reaction chemistry would, in three years, also earn a Nobel. Just a year older than I, Manfred was boyishly enthusiastic about an amazing range of nonchemical activities, especially the piano. At his home he assembled a small chamber orchestra to accompany him as he raced through a Mozart concerto with only occasional mistakes.

My main purpose in going to Europe was a June conference, "The Principles of Biomolecular Organization," at the CIBA House in London. It was effectively a follow-up to the "Nature of Viruses" meeting there nine years before. Just before the meeting, J. D. Bernal, the head of the Birkbeck College lab where Rosalind Franklin had moved after leaving King's, suffered a mild stroke that made his delivery of the opening remarks painful for those who had long known him as "Sage." No trace of a halting brain appeared in his later published piece stating that life is in no sense a metaphysical entity but a precisely patterned structure right down to the atomic level. Feodor Lynen was also in the audience and both of us were intrigued by the last talk, "The Minimum Size of Cells," by Yale's Harold Morowitz. Here he focused on the smallest free-living cells, organisms such as PPLO and mycoplasma, whose genomes likely contained fewer than a million base pairs.

I left central London after the meeting to visit Av and Lorna Mitchison on the Ridgeway, near the grounds of the Mill Hill Laboratory of the Medical Research Council. A number of family friends were there, the youngest being the intelligent and statuesque Susie Reeder, about to receive her university degree at Sussex and soon to commence a postgraduate degree program in criminology at Cambridge. The next evening we had dinner at Rule's Restaurant, just below Covent Garden. It was just a few minutes' walk away from the Aldwych Theatre, where we saw Harold Pinter's *The Homecoming,* with Ian Holm and Vivian Merchant, then Pinter's wife. Afterward, I walked Susie across

Waterloo Bridge, where she caught a train to her mother's home in Putney.

A month later Susie was to be in the States on her way to a month-long holiday near Denver, where she planned to visit her British boyfriend. She seemed eager to stop off in Cold Spring Harbor, where in mid-July I would be staying at the home of the lab's director, John Cairns. In the end she came only for a day, letting me admire her swimsuited form on the raft off the lab's beach. Ensuring that the occasion's memory would not be one to cherish was the continuous presence of the Cairnses' German police dog, who nearly bit me on the leg before being dispatched. Early in September, on her way back to England, Susie stopped off in Boston long enough to let me take her to supper at the Union Oyster House after I'd ruefully observed her lack of attention to the art on the walls of my Appian Way flat.

Two weeks before, W. A. Benjamin had held a party at Woods Hole to mark the appearance of the first copies of *The Molecular Biology of the Gene*. Many from my Harvard lab came down for the event, as did my dad, whom Neil Patterson kindly spent much time talking to. Earlier in the day, I had driven down with a pretty Radcliffe senior with short blond hair called Joshie Pashler, who also had something to celebrate in the recent discovery of her first RNA phage R17 mutant. Over the summer, I let Joshie read newly completed *Honest Jim* chapters, hoping each would make her want to read the next one. Like my previous intelligent and pretty assistants, Joshie had a steady Harvard boyfriend, though I noted by way of small comfort that she was a good deal sharper than he was.

Also at the W. A. Benjamin party was Keith Roberts, back for much of the summer. His first months up at Cambridge had not been what he anticipated. To start with, his longtime girlfriend from Norwich, Jenny, had taken up with "a more mature, better-looking man." And his Cambridge digs, while cheap, were grim—one small room without the typical fireplace. Moreover, he found life at his college dull, with everything closing down at 11:30 P.M. and a gown required while dining in St. John's five nights a week. As modest distractions, he took out subscriptions to the Arts Society, the Humanists, and the Marxist Society. He also had the pleasure of hearing Francis lecture to the

Humanists on the topic "Is Vitalism Dead?" with a fiery delivery that put off those not sharing the view that religion is a mistake. Lately things were looking up, however, as Jenny had come back to Keith before the academic year ended and was with him at the Woods Hole party. Without Keith's artwork, my text would not have sold almost twenty thousand copies the first year and even more the second, generating yearly royalty income soon equal to my Harvard salary.

During the coming academic year, I planned to be on sabbatical in Alfred Tissières's lab in Geneva, with half of my Harvard salary supplemented by a Guggenheim Fellowship. As the time to leave approached, though, I realized the first half of sabbatical time would be better spent finishing *Honest Jim* in Cambridge. There I could probe Francis's memory of key moments in our adventure. Before flying across the Atlantic, I sent my half-complete manuscript to Dorothy de Santillana to share with her Houghton Mifflin colleagues. Several weeks passed before I got a response from their chief editor, Paul Brooks, over a Friday lunch at his Boston club. In his opinion, the language was too strong and likely to lead to libel suits. But since he was not expert on such matters, he had arranged for me to visit Hale and Doer, the prestigious downtown Boston firm to which they had sent my chapters. Several days later, I again crossed the Charles River to go over several pages of comments by a Boston blueblood, improbably named Conrad Diesenhofer, on how my use of certain words would create trouble. I went back to Harvard suspecting that Houghton Mifflin's risk aversion could not allow them beyond the jeopardy that might attend issuing further editions of Roger Tory Peterson's bird guides.

In Cambridge, Sydney Brenner had arranged for me to have an absent fellow's digs at King's, overlooking the big green lawn that fronted Clare College. Once a chapter was finished, I took it to the Laboratory of Molecular Biology (LMB) for typing by Francis's young secretary. That I was writing up the story as a solo effort, not a joint project, at times seemed to annoy Francis. But on other occasions, he easily recalled key events that had already vanished from my memory. Later I showed my resulting efforts to Susie Reeder, whose postgraduate criminology studies were centered in a nineteenth-century house on West Road, close to where Lawrence and Alice Bragg had lived when I first arrived in Cambridge. On several evenings Susie and I drove to restaurants outside Cambridge using a rented MG sedan that also took me to and from the LMB.

During my last ten days in Cambridge, my father stayed nearby at the Royal Cambridge Hotel on Scroope Terrace, coming from central France, where he had been on holiday. Just before Christmas he flew back across the Atlantic to be with my sister in Washington. Simultaneously, I went up to Scotland to be at Dick and Naomi Mitchison's large house at Carradale. By then I had almost finished *Honest Jim,* having only two chapters to go. In Naomi's study, under one of her Wyndham Lewis drawings, I wrote out the next-to-last chapter. Once the New Year had been celebrated, I flew back to Harvard, where I needed only a day to compose the last chapter. The final sentence, "I was twenty-five and too old to be unusual," had been long in my mind.

The next day I spotted Ernst Mayr at the Faculty Club and told him that I had just finished my account of how the double helix was found. Then a syndic of Harvard University Press (HUP), Ernst quickly phoned Tom Wilson, the press's director, to inform him of my manuscript. The next morning, Tom came to my office and took away the manuscript for an overnight read, knowing I was returning soon to London. Excitedly he called the next morning to say he wanted the press to publish *Honest Jim.* With my permission, he would send copies to referees on the Harvard faculty, who helped decide which books the press should publish.

When I arrived in London, I immediately phoned my old friend Peter Pauling to suggest he bring along for a fun lunch the highly intelligent, suede-jacketed Louise Johnson. She, like Peter, was working on her Ph.D. at the Royal Institution, then presided over by Sir Lawrence Bragg. A month before, I had met Louise at a biophysics meeting at Queen Elizabeth College, learning that she worked on the structure of lysozyme with David Phillips. To my delight, Peter at the last moment got her and the equally young crystallographer Tony North to join us at Wheeler's Restaurant on Dover Street. Showing them my manuscript, I told them its novel-like construction necessarily led me in the opening chapters to portray Sir Lawrence in an unflattering way. I confessed now hesitating to show it to Sir Lawrence even though, more than a year before, he had suggested that I write my side of how the double helix was found. He had been worried at the time that Francis was getting too much of the credit for our big breakthrough. How to have him read my manuscript without infuriating him was a problem that stumped all of us until Tony's sudden brainstorm. Why not invite Sir Lawrence to write the book's foreword? By so doing, both he and I would come out on top.

I could not make that request of Sir Lawrence until March, after the conclusion of my six-week tour of East Africa, sponsored by the Ford Foundation. The trip was arranged at the last moment by Charlie Wyzanski, who, as a trustee, had recently been there looking at the foundation's African operations. The East African offices were in Nairobi, and I flew there in a BOAC VC-10, first class, as senior Ford Foundation staff traveled. Sitting next to me was a woman who asked to see the *Honest Jim* manuscript that I had by my side. This she read in two hours, telling me she could not put it down. After going on to give three lectures at Makerere University in Kampala, Uganda, Frank Sutton, the former Harvard junior fellow, joined me on a trip to see how foundation monies directed toward wildlife conservation were being used. First we stopped at Queen Elizabeth Park, on Lake Edward, above which the Mountain of the Moon rose into the clouds. Later, we were surrounded by hippos and crocodiles, while an old-fashioned, *African Queen*–like boat took us to the base of Murchison Falls, over which the White Nile cascades down.

Also staying at the Makerere Faculty Club was the Oxford-educated writer V. S. Naipaul, who, over several breakfasts, showed no interest in the fact that I too was generating English prose. Around the swimming pool, stand-offish in a quite different way, was the petite, well-shaped daughter of Donald Soper, the prominent English Methodist. Caroline betrayed no desire to learn from *Honest Jim* how Cambridge science moved. She was only there babysitting her sister's children while her Cambridge brother-in-law gave physiology lectures to Makerere students. They followed their Cambridge equivalents in wearing academic gowns in their dining hall, a custom I honored on several evenings, to my discomfort.

Upon the conclusion of my African lectures, which also took me to Tanzania, Sudan, and Ethiopia, I made Geneva my main base until my sabbatical ended in late May. From there I went twice to London, the first time to give Sir Lawrence my manuscript and tell him that Harvard University Press wanted to publish it. On my second visit to his Royal Institution top-floor flat, I waited nervously outside until upon entering he put me at ease, saying that by writing the foreword he could not sue me for libel. Later I learned that his initial reaction to my manuscript was, as feared, one of white-hot anger but that his wife, Alice, had cooled him down. Immediately I had Tom Wilson officially inform Bragg about HUP's publication plans. Within a month, Bragg delivered his crisp, elegant foreword saying I wrote with a Pepys-like frankness. Feeling that I had a book that I could now publish, I sent the manuscript to Francis asking for comments about its accuracy.

I also gave *Honest Jim* to my friend of many years Janet Stewart, then an editor at André Deutsch, to see whether they would like to consider it for British publication. Janet and I had first met in Cambridge when she was up at Girton College. Now married to the barrister Ben Whitaker, she had been in publishing in New York before returning to England as an editor with Deutsch. She and her husband lived on Chester Row and over dinner I gave them news of Bragg's foreword. Several days later I went to Deutsch's offices on Great Russell Street, where Janet introduced me to her Hungarian-born boss and uncomfortably looked on as he offered me a £250 advance. I politely replied I would pass his offer on to Harvard University Press director Tom Wilson. Later I told Tom to find a British publisher who understood that scientists are not indifferent to money.

Tom's choice proved also to be Hungarian, the portly, astute George Weidenfeld, said to be an inspiration for Kingsley Amis's novel *One Fat Englishman*. His publishing house, Weidenfeld and Nicolson, had recently gained notoriety by publishing *Lolita*, which had cost Nigel Nicolson his seat in the House of Commons (though it apparently did not interfere with Weidenfeld's being created a life peer in 1976). Tom sent *Honest Jim* to George in time to let him read it before I passed through London in mid-July on my way to a scientific meeting in Greece. Then Weidenfeld and Nicolson had offices above Bond Street. George himself lived in an Eaton Square flat whose high ceilings let him optimally display his large canvases, which soon included a Francis Bacon. Wasting few words, George offered me a $10,000 advance, half payable upon signing his contract, the other upon my book's publication. Immediately I accepted, asking him to send the contract to Tom Wilson to see that its terms were compatible with those in the contract I was about to sign with HUP.

By then Francis had let me know he did not like the title *Honest Jim*, its implication to him being that I alone was hawking the gospel truth. That one would hardly make such an assumption buying a used car from "Honest Francis," or even from "Honest Jesus," did not budge him. So I changed the title to *Base Pairs,* aware that this pun by itself might lead to a libel suit caused by a jacket cover with Francis and me and of Maurice and Rosalind staring at each other à la *Kind Hearts and Coronets.* Though Tom Wilson wanted me to use my third choice, *The Double Helix,* he went along with *Base Pairs* on the title page of an only mildly revised manuscript sent to Francis, Maurice, John Kendrew, and Peter Pauling, together with forms to sign saying they had no objections to HUP's publishing my manuscript.

We should have realized that neither Francis nor Maurice would see any advantage in giving permission to publish material that they felt not only failed to serve the public interest but also harmed them. Too soon their response came through a high-powered attorney, who wrote President Pusey of his clients' contention that they were libeled by my manuscript. They employed the same lawyer that Jacqueline Kennedy had unsuccessfully used to block publication of *The Death of a President,* William Manchester's book on her husband's assassination. Both Tom Wilson and I noted that this New York hired gun was careful not to specify that *he* considered my book libelous. Nevertheless, we feared that President Pusey would feel himself involved in something untoward, and Tom later unilaterally decided that HUP would move ahead only with the Harvard president's approval.

I also should have realized that Peter Pauling would feel a filial duty to send my manuscript to his father. After reading it, Linus fired off an angry letter to Tom Wilson calling *Base Pairs* "a disgraceful example of malevolence and egocentricity." He wrote demanding that I remove the lines "Linus's screwy chemistry" and "Linus looking like an ass." These were phrases I knew good taste would lead me to delete before the manuscript went to the printer. But since they were true, I was loath to remove them before absolutely necessary. In a similar vein, I never should have sent out a manuscript saying that Francis had never been a member of any college because he was thought to pinch other people's ideas. By this I only intended to convey the reason why King's, to their great loss, had not made Francis a fellow despite his unquestioned brilliance.

Already Tom Wilson had assigned his editor Joyce Leibowitz the task of working with me to make the manuscript less objectionable to the story's principals, at the same time preserving its aim to tell what really happened. Joyce had the wit to see that my story would benefit from an epilogue saying that my descriptions of Rosalind Franklin did not do justice to her scientific accomplishments while at King's. Also helping me was the bright literature major Libby Aldrich, who, like Dolly Garter, had taken George Wald's natural science biology course. Libby then was writing her senior thesis on Sylvia Plath, whom I remembered scurrying along King's Parade in Cambridge in the mid-1950s.

Soon after receiving the *Base Pairs* manuscript, John Kendrew took it to J. D. Bernal, who wrote him saying, "I could not put it down. . . . Considered as a novel of the history of science as it should be written, it is unequaled. It is as exciting as *Martin Arrowsmith.*" While asserting his opinion that I was unfair to the contributions of Rosalind Franklin and noting that I did not even mention the work of Sven Furberg, both from Bernal's lab, he offered the following comment: "Watson and Crick did a magnificent job, but in the process were forced to make enormous mistakes which they had the skill to correct in time. The whole thing is a disgraceful exposé of the stupidity of great scientists' discoveries. My verdict would be the lines of Hilaire Belloc:

And is it True? It is not True.
And if it were it wouldn't do."

I was also encouraged by a letter from the Hungarian-born immu-
nologist George Klein. During his late fall visit to Harvard Medical
School, my father prepared Sunday lunch at 10½ Appian Way for the
three of us. George left with a copy of *Base Pairs* to read on the plane
back to Stockholm. From the Karolinska Institutet he sent a letter saying that I had written "an unparalleled description of the excitements,
the frustration, the greatness, and the smallness of creative research
students. . . . You should expect a hostile backlash from most scientists; you should not try to soften your book, have it printed as it is or
not at all."

Over three winter months, I continually made minor changes to
correct misstatements of fact or personal intent, particularly as
enlightened by Francis and Sir Lawrence Bragg. Again using *Honest
Jim* as the title, I worried that Francis and Maurice's objections would
lead Sir Lawrence to withdraw his foreword, and I wrote to him that I
would understand if he saw fit to do so. On April 19 he replied that he
would be very sorry if it came to that, but that he also wished to state
categorically that his contribution was contingent on my making certain changes, in particular, one correcting my misstatement that
Perutz and Kendrew told him they would leave Cavendish if Crick was
fired. In conclusion, he "wished the book every success." At the same
time, John Maddox, the editor of *Nature,* found nothing libelous in
the newest *Honest Jim.* In fact, he believed it to be much less dodgy
than earlier reported to him: "In other words, I would like to see it
published." But when that occurs, he said, "you will have to barricade
yourself in for six months or so."

Both Maurice and Francis continued to oppose *Honest Jim* in every
way possible. In a letter, Maurice reminded me of having written when
I sent him my first draft, "You might think you have reason to shoot
me." Now Maurice worried that "his letter would make me want to
shoot him." In it he suggested that I abandon any thought of publishing my book intact but instead have its science passages incorporated
in the forthcoming book of historian Robert Olby on the double helix.

Contributions emerging from tape recordings of himself, Francis, Erwin Chargaff, and Pauling were also expected to appear in Olby's book.

Francis's five-page letter of April 13 started out saying that the new version was a little better but his basic objections were the same: "Your book is not good history"; "You did not document your assertions (with appropriate references) . . . displaying the history of science as gossip"; "Your view of the history of science is found in the lower class of women's magazines"; "If instead considered as autobiography, it is misleading and in bad taste"; "The fact a man is well known does not excuse his friends from respecting his privacy while he is alive"; "The only exception should be when private matters are of direct public concern like with Mrs. Simpson and King Edward"; "Your book is vulgar popularization which is indefensible." On the next-to-last page, Francis raised the stakes. A psychiatrist to whom he gave the manuscript reportedly said, "The book could only be made by a man who hates women." Another shrink concluded that I loved my sister to excess, "a fact much discussed by your friends while you were working in Cambridge, but so far they have refrained from writing about." On the last page, Francis noted copies of his letter were being sent to, among others, Bragg, Pauling, and Pusey.

A month later, Nathan Pusey told Harvard University Press that it could not publish my book, saying, "Harvard did not want to be involved in fights between scientists." Libel considerations were likely not involved but no foundation for such objections seems to have been established in any case. Some months before, at a party dominated by Harvard Law students, a recent law school graduate told me that he was reading my manuscript for Ropes and Grey, Harvard's Boston lawyers, explaining that he had no past experience with libel matters. Once I learned that HUP was out, Joyce Leibowitz suggested I retain as my personal counsel the New York lawyer Ephraim London. A greatly respected legal scholar, Eph had successfully argued several freedom-of-speech cases before the United States Supreme Court. A tall, thin man with connections to publishing going back several decades, Eph had been Simon and Schuster's house counsel. Upon reading my manuscript, he said it contained no libel.

Already I had a new publisher, the newly formed Atheneum Press, started by Pat Knopf, son of the famous publisher Alfred A. Knopf, and Simon Michael Bessie. I chose it to stay with Tom Wilson, who was resigning as director of Harvard University Press to join Atheneum. Tom's leaving HUP had nothing to do with Pusey's decision to block my book. The decision had been made beforehand, in response to his approaching HUP's compulsory retirement age. Tom's children were still young and he needed a well-paid job into the foreseeable future. Ironically, only because HUP was not publishing my book could I continue to enjoy the reassurance of having Tom at my side. Lawrence Bragg, confident of Tom's integrity, let his preface stand. Fearing a libel action if not a petition for an injunction against *Honest Jim*'s publication, Atheneum retained the New York lawyer Alan Schwartz, whom William Manchester had used to defend against Jacqueline Kennedy's libel suit.

I met with him and Tom Wilson several times before Eph London came into the picture to tell Schwartz that the changes he wanted were unnecessary, as *Honest Jim* was neither libelous nor an unwarranted invasion of privacy. Many of Schwartz's suggestions would blunt intended candor. His accepted version of the first sentence, "I can't ever remember seeing Francis Crick in a modest mood," could only have been written by a timid lawyer. In a few cases, he wanted harmless substitutions such as *often* instead of *generally*. To these I gave in. As Tom Wilson strongly concurred, I also agreed to Schwartz's wish that the title become *The Double Helix*.

The situation was less under control across the Atlantic, where Weidenfeld's solicitor, Colin Madie, still maintained that Francis was being defamed and that, given his reputation for not being particularly well balanced, we should not expect him to act in his own long-term interest. By the end of September, Madie abruptly changed his opinion, telling Weidenfeld to proceed. This happened after he showed the manuscript to a close friend who had known Francis for years and who told him that my portrait of Crick was "right on mark." Weidenfeld's editor in chief, Nicolas Thompson, then reread the manuscript, writing to me that "my picture of Francis was one only a hypersensitive or very unreasonable person could object to. You point to his faults indeed, but much more to his enormous talents and likable qualities."

The way was now clear to sign final contracts, with Tom Wilson looking embarrassed as he conveyed to me Simon Michael Bessie's offer of an André Deutsch–magnitude advance. Seeing no gain from asking Bessie why he took me for a fool, I let Eph negotiate a more reasonable sum. Later, Bessie tried to renege on his promise to stipulate that Atheneum would pay one-half of any costs of successfully defending a libel suit. So I wrote him that the contract already incorporated all compromises, and there we must stand. Otherwise, I would find another publisher, notwithstanding my connection with Tom. I gave him a deadline to back down, which he did. I felt sorry for Tom's having to be associated with this overrated publisher, who was not a patch on someone like George Weidenfeld.

A better side of Atheneum was presented by Harry Ford, who chose the typeface and designed a striking red jacket. Once I got through my first fall Harvard lectures, I assembled and sent on to him the appropriate photos and preliminary sketches for diagrams of DNA bases, the sugar phosphate backbone, and so on. Libby Aldrich was no longer available to help me, having gone off to Lady Margaret Hall at Oxford to study English and also to avoid facing her emotions concerning the *Advocate's* former editor Stuart Arrowsmith Davis. In Plath-like fashion, she wrote a blue-tinted letter to describe herself as freezing, pale, and gaunt, but very well acclimated to dropping shillings into various heating devices and visiting the public baths (another shilling) along with the rest of the neighborhood's female population of Indians and Cypriots. Lady Margaret Hall itself, she said, was part convent, part prison, and very much an autonomous little private girls' school, through which passed innumerable withered little old ladies, two of them her tutors—the Anglo-Saxon one old and fierce, the literature one old and sadly girlish. For cheer, pictures of Mick Jagger and Bob Dylan were on the walls of her two attic rooms, above the floors occupied by her meek Irish landlord and his virago wife. Seeing *Privilege,* Libby wrote that she shortened her skirts and aimed to cultivate glamour.

Tom Wilson now was in a position to contact the *New Yorker* about serializing *The Double Helix* as they had Truman Capote's *In Cold Blood.* But they turned us down, as did *Life* magazine, which said they had already dealt with DNA through their big 1963 illustrated article. *The Atlantic Monthly* responded more positively, publishing *The Double Helix* intact in their January and February 1968 issues. By then, Francis and Maurice had given up thoughts of any legal action, with Francis feeling victorious over HUP's withdrawal. No one would now have cause to think *The Double Helix* a scholarly book. Early in February 1968, Eph sent me a bill for $700 for his assistance between June 1967 and October 6, 1967, citing charges for (1) his opinions with respect to libel, (2) the withdrawal of Atheneum-suggested changes that he thought unwarranted, and (3) conferring with attorneys for Atheneum with respect to requests by Dr. Crick's attorney to examine manuscripts, correspondences, et cetera. To complete the bill, $8.86 was requested for toll calls and $5.75 for messenger service.

A luncheon was held at the Century Association on February 14, 1968, for reviewers and science editors. There I would have to be gracious to Michael Bessie, but Libby Aldrich was on hand for me to make snide remarks to behind his back. Just before Christmas, Oxford had dropped out of her life, and for six weeks she'd expected to be Mrs. Stuart Arrowsmith Davis. The wedding was meant to take place in Bronxville the Saturday before my event. Just before the ceremony, however, the groom had suffered a nervous collapse and the marriage was indefinitely postponed. By the luncheon's end, Libby was nowhere to be seen, and I eventually found her in a ladies' room passed out from having drowned her sorrows in pre-luncheon drinks. A cab took us to the Plaza, where I had a big room with a window on the park. Libby instantly fell asleep in my bed. By 7:00 P.M., she was alert enough for dinner at La Côte Basque before I took her to Grand Central Station for the train to New Haven, where Stuart was a Yale graduate student.

The next evening I met the Atheneum publicity agent at the studio where I was to appear on Merv Griffin's TV show. My conversation with Griffin seemed to end almost before it started, with my nervous movements causing Merv's English-butler sidekick, Arthur Treacher (also of eponymous fish and chips fame), to ask whether I needed the little boys' room. Ten days later, I went back to New York to appear after Harry Belafonte on the *Today* show and attend a luncheon to mark the book's official publication date. In the middle of March I was there yet again for a book world luncheon at the Waldorf-Astoria. By then, several positive reviews had appeared, the most important by the Columbia University sociologist Robert Merton. The article, entitled "Making It Scientifically," began: "This is a candid self-portrayal of the scientist as a young man in a hurry." Richard Lewontin used his *Chicago Sun-Times* space to compare it to Françoise Gilot's *Life with Picasso,* calling it a vulgar curiosity about minor scientific celebrities. Soon I was on the *New York Times* best-seller list, remaining there for sixteen weeks, though never near the top. *Time* magazine for some two weeks wanted me on its cover, sending a reporter to follow me about at Harvard and then watch me speak at Dartmouth. Eagerly I sought out *Time* on the day promoted for my front-page appearance only to see the face of "Danny the Red." The student barricades in Paris had become more important than DNA.

Only late in May did Weidenfeld publish the British edition of *The Double Helix.* They had produced a much condensed version for the *Sunday Times* to publish, but I nixed the effort, saying that it lacked the character of the full book and would unnecessarily annoy Francis and Maurice. Even worse was the vulgar jacket, printed a month before publication without my input. It made Francis seem ridiculous, with such ludicrous attempts at seductive copy as: "1) Which winner of the Nobel Prize has a voice so loud it can actually produce a buzzing in the ears? 2) Who is the top Cambridge scientist who gossips over dinner about the private lives of women undergraduates? 3) Which eminent English biologist created a scandal at a costume party by dressing up as George Bernard Shaw and kissing all the girls behind the anonymity of a scraggy red beard?" Mortified by my publisher's stupidity and grossness, I immediately contacted Nicolas Thompson to have Weidenfeld replace the offensive jacket. With no argument they backed down, and George Weidenfeld personally reassured me that all the jackets were being destroyed.

The week I was in England to mark the official publication date was no time to try to see Francis and Maurice. But Peter Pauling was typically fun to be with, and I could deliver the English version to Naomi Mitchison, to whom I dedicated *The Double Helix*. I saw Lawrence and Alice Bragg at their country home near the Suffolk coast. In England, most of the reviews were favorable. The most critical was by the embryologist C. H. Waddington, who thought me verging toward Salvador Dali–like manic egocentricity. By the year's end, some seventy thousand books had sold in the United States and British sales approached thirty thousand. Given its generally high praise and wide visibility, Tom Wilson thought I would be a shoo-in for the 1969 National Book Award in science. But it went to Yale's Robert Jay Lifton for *Death in Life: Survivors of Hiroshima*.

Though I was disappointed, I no longer needed others to tell me I had written a book worth reading.

Remembered Lessons

1. Be the first to tell a good story

In 1953 the finding of the double helix by itself did not create the opportunity for an important new textbook. Any such book written the next year necessarily would have been dominated by other facts already well documented—and which still constituted most of what was known on the subject of life's nature. Twelve years had to pass before an almost complete, new story could be told of how the genetic information within DNA molecules is used by cells to order the amino acids in proteins. By contrast, the story of the quest for DNA's structure could be told immediately, although it took me almost a decade to figure out how to go about telling it. Many have had their objections to my version of characters and events, but the popular imagination was captured by it, not least on account of its having come first.

2. A wise editor matters more than a big advance

Assuming you are not being insultingly low-balled, choosing a publisher on the basis of the advance is like choosing a house builder solely on the basis of the lowest bid. An innovative book usually takes more time to write and may cost more money to produce than either you or your publisher would guess at the time of signing the contract. Better to have a seasoned and comprehending editor on your side when your manuscript takes many more years to finish than contractually stipulated. By then your editor, if not employed elsewhere, will be under pressure to curb production costs as much as possible. Your illustrations may be cut in number and fobbed off on the cheapest available commercial artist, but the chances of that are diminished if the publisher isn't already deep in the hole having paid you money you haven't yet earned. If you haven't been overpaid, your freedom to pay back the advance and take the book elsewhere is greater and so is your leverage in demanding that corners not be cut.

3. Find an agent whose advice you will follow

Publishers' contracts invariably contain clauses that only publishing lawyers understand. Unless you want to become credentialed in this arcane specialty, another field that has seen its best days, let your prospective contract go through the hands of someone paid by you to see that you are not taken advantage of. It is too much to expect your publisher, no matter his reputation for rectitude, to look after your interests and his own equally. The 10 percent to 15 percent of the proceeds charged by a reputable agent are well worth whatever is saved trying to represent yourself.

4. Use snappy sentences to open your chapters

With so much on TV, a short, incisive first sentence is more important than ever in pulling your readers into a new chapter. Let your audience know where they will be going if they stay with you. In *The Double Helix,* I used openers such as "I proceeded to forget Maurice, but not his DNA photograph." Equally important are ending sentences, in which I often sprinkled a touch of irony, as in "The remnants of Christianity were indeed useful," or attempted Oscar Wilde–like epigrams: "The message of my first meeting with the aristocracy was clear. I would not be invited back if I acted like everyone else."

5. Don't use autobiography to justify past actions or motivations

A major reason for writing autobiography is to prevent later biographers getting the basic facts of your life wrong. If life has graced you with lots of memorable occasions, merely reporting them correctly and dispassionately will generate a book worth reading. Attempts at justifying your actions and apologizing for bad behavior long ago only consign your work to the dubious genre of apologetics. Better to tell it straight without vainglory or shame and let others praise or damn you, as they will inevitably do anyway.

6. Avoid imprecise modifiers

Modifiers such as *very, much, largely,* and *possibly* don't convey useful information and only reduce the impact of otherwise crisp language. Saying someone is very bright offers no further insight than just saying he is bright. To go further, you must be more creative; for example, comparing your subject's brightness (or stupidity) with that of a known person or somehow ranking him, saying for instance, "No one was brighter in the Cavendish Laboratory"—that's got to mean something.

7. Always remember your intended reader

From the start, I wanted *The Double Helix* to be read beyond the world of science. So I integrated paragraphs about science with ones dealing with people, their individual actions and motives. Technical facts not essential to the story I left out. Even so, I found certain highly paid lawyers annoyed by any paragraph too technical for them to understand. I savored the justice of that.

8. Read out loud your written words

To make *The Double Helix* read smoothly, I read aloud every sentence to see if it made sense when spoken. Long sentences that were hard to follow I broke into shorter ones. I also sometimes combined a few short ones, as one short sentence after another can obscure the significance of events that unfold over more than one day. Choppy language is better suited for cookbooks and lab manuals.

BY THE mid-1960s, more and more of the research being done in Wally's and my third-floor labs was directed toward understanding how gene functioning is regulated by specific environmental triggers. We were preoccupied by concepts emanating over the past decade from the Institut Pasteur in Paris. There Jacques Monod and François Jacob skillfully employed genetic analysis of the bacterium *E. coli* to study how its exposure to the sugar lactose induced the preferential synthesis of the lactose-degrading enzyme β-galactosidase. They showed the existence of a lactose "repressor" whose presence negatively controls the rate at which β-galactosidase molecules are made. Their work suggested that free lactose repressors bind to one or more regulatory regions on the β-galactosidase gene, thereby preventing subsequent binding of the RNA-making enzyme RNA polymerase. In their 1961 Cold Spring Harbor Symposium paper, Jacob and Monod had proposed that the lactose repressor was an RNA molecule. Controversy by now existed as to whether they were correct, with others suspecting it to be a protein.

In 1965, Wally's main aim was to isolate the lactose repressor. As it was likely present only in a few molecules per bacterial cell, its identification was not a task for the faint-hearted. Two years before, Wally had spent several months unsuccessfully searching for it, believing it should specifically bind to β-galactosidase inducers. Sensing then he was going nowhere, he turned to experiments with Julian Davies and Luigi Gorini that revealed streptomycin-induced misreadings of the

genetic code, which offered possible explanations of how this powerful antibiotic kills bacteria.

Also keen to get the lactose repressor was the German biochemist
Benno Müller-Hill, who was one year younger than Wally. Coming
from a politically liberal family, Benno gravitated further to the left as
a chemistry student in the German socialist student scene, discovering
that many teachers at the University of Munich had been Nazi sympa-
thizers, though no one in authority seemed to care. In his hometown
of Freiburg, Benno later did doctoral work in the laboratory of the
sugar chemist Kurt Wallenfels. There he learned the essentials of protein chemistry through studying β-galactosidase. He began to love science and became excited about Monod and Jacob's work on how
lactose molecules induce the synthesis of β-galactosidase. Later, in the
fall of 1963, Benno began a postdoctoral position in Howard Rickenberg's lab at Indiana University, to which he brought samples of the
Wallenfels lab's glycosides to study their specificity in inducing
β-galactosidase.

In Bloomington, Benno never felt comfortable either as a German
among its many Jewish biochemists or as a leftist among Americans
whose paranoia about communists in their midst surprised him. But
his experiments there, which gave him sufficient results for a talk at
the August 1964 International Biochemistry Congress in New York,
were an ample reward for such social unease. By then he wanted to
move on to the lactose repressor and came up to me after my talk to
see whether I would accept him into my Harvard lab. Explaining that
it was Gilbert he needed to approach, I urged him to visit Harvard as
soon as Wally returned from a lengthy visit to England. Upon their
meeting, Wally instantly saw Benno as the collaborator he needed and
offered him a research position starting as soon as he could politely
leave Howard Rickenberg's lab.

In Bloomington, Benno had learned how to genetically manipulate
E. coli. He ably deployed this newly acquired skill soon after he arrived
at Harvard to show that the lactose repressor is indeed a protein, not
an RNA molecule. By using chemical mutagens he generated almost
two hundred *E. coli* mutants that made β-galactosidase in the absence
of any inducers. Two of the mutants represented change to "nonsense"

codons leading to premature polypeptide chain termination. If the repressor was made of RNA, this class of mutants would not have existed. The simple elegance of Benno's experiment was not revealed in his first manuscript draft. After telling him it was heavy and Teutonic, I rewrote it before its October 1965 submission to the *Journal of Molecular Biology*. As one of the journal's editors, I knew the article would quickly appear in print.

Though both Benno and Wally had earlier independently failed to detect the lac repressor through its binding to potent β-galactosidase inducers, this feature offered still the only approach at their disposal. To increase their chances of succeeding, Benno again turned to bacterial genetics, making a mutant repressor that had enhanced affinity for the chemical that induced isopropyl-β-D-1-thiogalactosidase (IPTG). By growing *E. coli* cells in very low concentrations of IPTG, a much more effective repressor became available. To double repressor numbers in bacteria, Benno made a diploid derivative containing two copies of its respective gene. These genetic tricks by themselves, however, were not sufficient to pinpoint the lac repressor in cell-free bacteria extracts. Success came only through developing molecular separation procedures that yielded protein samples enriched in the lac repressor. The first positive results were achieved in May 1966, but they were barely credible. Only 4 percent more radioactively labeled IPTG was found in bacterial extracts containing repressors than in surrounding repressor-free solutions. Soon better fractionation methods led to a semipurified sample that drew the IPTG into a semipermeable dialysis sac at a concentration almost twice that found outside. These enriched extracts were not affected by the enzymes that break down DNA and RNA. In contrast, the protein-degrading enzyme pronase destroyed all binding activity, confirming Benno's genetic pinpointing of the repressor as a protein.

Until then, Wally and Benno faced the likely prospect of not being first to characterize a repressor's molecular nature. On the fourth floor was twenty-six-year-old Mark Ptashne, who was feverishly trying to isolate the phage λ repressor. It blocks the functioning of all but one phage λ gene when phage is present as an inactive prophage on an *E. coli* chromosome. The only λ gene then functioning is that coding for the repressor. Though its existence became known through elegant genetic experiments at the Institut Pasteur, no one in Paris had come up with a workable approach for its molecular characterization.

Mark had arrived in the fall of 1960 to do his Ph.D. thesis work with Matt Meselson. As essential to his nature as his desire to do top science were his leather motorcycle jacket, his violin, and his golf clubs. In high school, Mark had spent summer vacations at the University of Minnesota working in the neurophysiology lab of a left-wing family friend. At Reed College, which he chose over Harvard for its exclusive devotion to undergraduate education, he moved from philosophy to biology, working during the summer before his senior year at the University of Oregon. There Frank Stahl told him to do his graduate work with Matt Meselson. Mark already knew that the λ repressor was the next big objective in the phage world. But this goal was too risky for an early 1960s Ph.D. thesis, and so Mark settled in for a semiroutine genetic analysis of phage λ. As his thesis experiments neared their end, Paul Doty and I strongly backed his appointment to a three-year stint in Harvard's Society of Fellows. This would give him a shot at the λ repressor. As a candidate, he proved a shoo-in, since Wassily Leontief, the new head of the Society of Fellows, saw in Mark an agreeable conversationalist for the society's Monday night dinners. His term as a junior fellow commenced in July 1965.

That August, I submitted a $55,000 grant application to the National Science Foundation to pay Mark's salary and lab expenses for the three years, including a $5,000 yearly salary for a technician. The funding would allow him to work independently from his boss, Matt Meselson, who by now had despaired of Mark's sometimes sloppy work habits. In fact, my application stated that Mark intended to use DNA-RNA hybridization techniques to detect the λ repressor, which was yielding messy results even before the grant came through. Not enough was known about how RNA polymerase transcribes genes in cell-free extracts.

Mark's game plan soon changed. He began to look for differences in the proteins synthesized when heavily irradiated bacteria are infected with different types of λ phage. He guessed that λ repressor synthesis constituted only 0.01 percent of the protein synthesis in cells carrying

Mark Ptashne lobs a softball at the 1968
Cold Spring Harbor symposium.

With Wally Gilbert's and my students and postdocs
on the Biology Department rhino in 1965

λ prophages. To make scarce repressor molecules visible, he needed to drastically reduce synthesis of most bacterial proteins, as well as to inhibit the synthesis of all λ-specific proteins that were not the repressor. He reasoned he could cut back the routine synthesis of cellular proteins by irradiating the bacterial host cells with massive doses of ultraviolet light.

Though Mark's experimental design was elegant, making it work would be no cakewalk. Though he received hints of early success, these were cruelly followed by failures to spot a radioactively labeled protein. In the summer of 1966, virtually all of Mark's experiments were crashing while Wally and Benno provided mounting proof that they were looking at the lac repressor. Happily, Mark's world would brighten immeasurably through the sudden unanticipated arrival of my former Radcliffe tutee, Nancy Haven Doe.

Nancy had been intrigued by repressors ever since learning about them during my spring 1963 Biology 2 lectures. Until then, she had expected her life to be largely that of the wife of a social male, very likely Brook Hopkins, Harvard '63, whom she had met as a freshman and with whom she had persevered through five years of "understood engagement." During her senior year, noticing Nancy's intellectual vitality, I strongly encouraged her to go to graduate school. Aiming her toward the best, I wrote to Rockefeller University's president, Detlev Bronk, in support of her admission. Perhaps because she had come to science so recently, Bronk did not take the bait. Nancy's fate instead became Yale, possibly pushed ahead by my recommendation letter describing her as a quick learner who happened also to be cheerful and pretty.

Nancy's first year in New Haven was a typical full load of four courses during both the fall and spring terms. She mastered the Schrödinger wave equation as well as many facts of chemistry that with luck she would never need to use. Eight straight A's left Alan Garen no choice but to accept her into his molecular genetics lab, where she wanted to go for the repressor. Soon, however, she realized Alan to be a man of few words, little time for mentoring, and an excess of caution. He told her he was not up to the repressor; it was too hard a problem for someone over thirty-five. The dull alternative he proposed held little hope of sustaining and exciting her as a scientist. Writing to me early in March 1966, she remained resolute about resisting contentment in mediocrity. By late spring, she could take no more of her New Haven abode and decamped with an equally disenchanted aspiring female academic to the island of Mykonos.

Upon coming back to the States, she feared that staying at Yale would condemn her to work in Bill Konigsberg's lab on dull, dull hemoglobin. In August, she wrote to me proposing to join Mark Ptashne's lab as his technician. There her three-and-a-half-year-long obsession to work on repressors could find a proper outlet. Only several days before, she'd visited Harvard and found Mark so clearly in need of intelligent help that he would forgive her several blessings of heredity including a six-inch advantage in height. Nancy found it infinitely more gratifying to work as a technician on the repressor than to be a graduate student not working on the repressor. Wanting my opinion about her potential new career, she made it clear she had not at any time been, nor ever would be, in love with Mark Ptashne. With that reassurance, I gave her my blessing.

Nancy nevertheless proved just the tonic Mark needed. With her methodically meticulous presence ensuring that he did not leave out essential experimental reagents, Mark could consistently detect the λ repressor and begin its molecular characterization. They soon showed that it was a protein of molecular weight near 30,000. Though Nancy at first believed Mark could win the repressor race, he knew otherwise. Wally and Benno were already writing up their paper, while he likely needed at least six more weeks in the lab before starting to write. Benno regularly came up to the fourth floor to check on their progress, much to Nancy's annoyance. She perceived his main purpose as gloating. Seeing Mark's always polite reception of Benno, however, Nancy managed an equal courage and grace. As for Wally, Mark so revered him that losing the race to him could never be devastating. Nancy's huge respect for Mark was no less evident: every day she went to the sandwich truck on Divinity Avenue to get his lunch, an egg salad sandwich, accompanied often by a chocolate éclair to fortify his morale.

Wally and Benno's paper "Isolation of the Lac Repressor" was submitted by me to the *Proceedings of the National Academy of Science*

(PNAS) on October 24, 1966, just in time for publication in the December issue. If I had delayed its submission to let Mark complete the experiments needed for his paper, Wally and Benno's discovery would bear the next year's publication date. Mark would have preferred this, but I argued that no one would deem his work less important for appearing in print second. On December 27, "Isolation of the λ Phage Repressor" went off to *PNAS*, to appear in the February 1967 issue. Mark triumphantly announced the λ repressor isolation at a seminar in our lab's tearoom a month before the paper came out. The room was packed and during his moment of triumph ("I did it all alone!") he never acknowledged Nancy's key role in his success. Afterward, I badly chewed him out. Instantly realizing he had been too full of himself, that evening he called Nancy to apologize and later sent her flowers.

Mark and Nancy went on to test whether their repressor worked by binding to specific DNA sequences. When mixtures of radioactive λ repressor and λ DNA were centrifuged together in sucrose gradients they sedimented together. In contrast, mixtures of λ repressor and DNA from phage λ imm434 did not co-sediment. These much hoped for results were nonreproducible for an awful week until Mark realized that Nancy had inadvertently raised the salt levels in their mixtures. She nervously repeated the experiment using the original salt levels. The new results were just coming off the isotope counter when Nancy and Mark had to attend a seminar on the floor below. Halfway through the talk, the suspense overwhelmed her and she returned to the lab, where she quickly realized all was again well. She and Mark were for the first time ahead of Wally and Benno. Returning from the seminar, Mark shouted for joy, and they went into the halls to spread their good news. Going downstairs, they caught Wally and me about to leave the Biolabs. Upon learning that the experiment was repeatable, Wally's face turned ashen. Mark's overtaking him was not acceptable.

Over the weekend, Wally set about to do the analogous experiment using DNA from a phage Benno obtained from Jon Beckwith of Harvard Medical School that carried the β-galactosidase gene and its control region. By Monday morning, Wally let it be known that he had preliminary positive results he intended to verify quickly, bringing him again even with Mark and Nancy. Seeing Wally's intense competitiveness made Nancy feel as if she could never survive the male-dominated dog-eat-dog grind of being a scientist. It also was an eye-opener for Mark. His reverence for Wally had taken a big knock.

But Wally's expectation of catching up to Mark and Nancy hit some skids. Holding him up were difficulties in purifying the radioactively labeled lac repressor. Though Mark gallantly held up his submission to *Nature,* intending to let Wally and Benno publish simultaneously, after two months they were not ready even to start writing, so he sent his paper in without theirs. Having been forewarned about Mark's submission, *Nature*'s editor, John Maddox, sent it to the printer the day it arrived. "Specific Binding of λ Phage Repressor to λ DNA" appeared on April 15, 1967, only six days after arriving in *Nature*'s London office. Wally and Benno's paper "The Lac Operator Is DNA" was submitted on October 28 to *PNAS,* in time at least for publication in the same calendar year.

Nancy became a graduate student again in the fall of 1967 and continued to work in Mark's lab. I had told her to apply to Harvard and a week later she was formally accepted. Mark also encouraged her, saying she was brighter than any of the relatively few women and more than half the men at Harvard. She glowed in the knowledge of the approval of someone she revered. After the misery of Yale, everything had worked out, including her engagement to Brook: soon after becoming Mark's technician she had also become Nancy Hopkins. They moved to Trowbridge Street, their apartment graced by my wedding present of a life-size leather pig, similar to one Brook had long admired in his exclusive "final club" on Massachusetts Avenue.

After his second paper came out, Mark devoted more time to being a visible leftist, joining several South American neurophysiologists in Havana, where he met the cartoonist Jules Pfeiffer. Later he would go to Hanoi and Saigon on a trip organized by a former junior fellow, the MIT linguist Noam Chomsky. Over the past year, our tearoom conversations had centered more on Vietnam than on science, and I remained glued to an issue of the *New York Times* covering the Tet offensive all through a talk by MIT's Ethan Signer on bacterial-gene-carrying λ phages.

Earlier Benno and Jon Beckwith had marched down Massachusetts Avenue from Cambridge to Boston to demonstrate against America's Southeast Asia policy, discussing λ phages along the route. Immediately following Tet, Benno went with his soon-to-be wife, Barbara, to New York City for a big antiwar rally in Central Park. Equally anti-establishment in outlook, Barbara had first met Benno at one of Mark's parties in his flat north of the Biolabs, on Sacramento Street. Mark's blond-haired, live-in girlfriend, Micky, was from an even more leftist, if not Stalinist, background. Mark frequently escaped with his violin to their cottage on Cape Cod for weekends of heavy fellow-traveler chitchat.

In the summer of 1967, Wally and Celia Gilbert's three-year-old daughter, Elsbeth, was diagnosed with an incurable metastatic sarcoma. They made constant visits to Children's Hospital, compounding Wally's exhaustion when he got back home from late evenings in the lab. His two other children, John and Kate, earlier had thrilled to their father's eureka at finding the repressor. Now the mood of the house on Upland Avenue was grim. Hoping to distract them, in mid-fall I persuaded an ABC-TV producer, Ernie Pendrell, to feature Wally and Mark on a forthcoming TV special about science. Initially he had wanted to do an hourlong feature about my work, sponsored by North American Rockwell. But I told him the race for the repressor would show science at its brightest, making the public aware that ambitious young scientists, like young poets, are more creative when not encumbered by the braking powers of maturity, which I already felt nipping at my own heels. With Harvard's approval, an ABC crew came during the second week of February 1968 and followed me and Wally walking in Harvard Yard with Wally's children eagerly asking on cue, "Did you find it yet, Daddy?" A week later, the crew returned for a raucous Lincoln's Birthday party I threw in my Appian Way flat.

By then, Mark was secure in the knowledge that as of July 1, he would be a member of the Harvard faculty. By appointing him a lecturer in biochemistry instead of an assistant professor, the University could match the salary recently offered him to be an associate professor at Berkeley. In support of his appointment, François Jacob wrote to Paul Doty that "Mark is the most gifted young man in his

```
                        September Noon Seminars


   19 September, Tuesday        On the Release of Formate from Nascent Protein
                                JERRY ADAMS


   20 September, Wednesday      Origin of Chloramphenicol Particle Protein
                                BOB SCHLIEF


   21 September, Thursday       Early Functions of Bacteriophage λ
                                ROGER HENDRIX


   22 September, Friday         Specific Nucleotide Accepting Activity of R17 RNA
                                BOB KAMEN


   26 September, Tuesday        The R17 A-Protein
                                JOAN STEITZ


   27 September, Wednesday      The Reconstitution of Infective R17
                                JEFF ROBERTS

   28
   28 September, Thursday       Methionine Aminopeptidases in E. coli
                                VOLKER VOGT




   At noon.                                            Room D-375
                                                       Biological Labs
   September 1967.                                     Harvard University
```

Noon seminars from September 1967

generation of biologists." Franklin Ford orchestrated swift movement by the administration. The day after Christmas he wrote to tell me Ptashne was confirmed in his decision to stay at Harvard. Several months after, when Mark's appointment formally began, the Harvard Council of Deans honored him and Wally with the prestigious Ledlie Award for 1968, which included a $1,600 honorarium.

*Frank Stahl (back to camera),
Gerald Selzer, Mark Ptashne,
Wally Gilbert, and Celia Gilbert
at the 1968 Cold Spring Harbor
symposium*

*Wally Gilbert and Max
Gottesman at Cold Spring
Harbor in 1968*

By then Benno had returned to Germany as a professor at the Genetics Institute of the University of Cologne. Before going, he again made elegant use of *E. coli* genetics to generate a mutant that overproduced the lac repressor. Putting this mutant gene on a λ-like phage led to hundredfold amounts of the typical lac repressor output. Just before Benno left Harvard, Jacques Monod was visiting and popped his head in to say hello and comment on Benno's identification of the lac repressor, remarking, "After all, Benno, it was pedestrian." Then he turned around and left. It was Jacques's sour grapes way of admitting that clever experiments, not a clever idea, had beaten him and his friends at the Institut Pasteur. With Benno, Wally, and Mark's success, the center of gene regulation was now not Paris but Harvard.

An unanticipated by-product of Wally's lac repressor purification effort was helping my graduate student Dick Burgess to optimize his RNA polymerase purification methodologies. More than ever, our lab wanted further insight into how RNA is made from DNA templates. In 1963, I'd had my graduate student John Richardson focus his Ph.D. research on RNA polymerase, the enzyme responsible for making RNA off DNA. RNA polymerase was discovered in 1959, but its molecular form was not easy to pin down. John found that the molecular weight decreases by half when the molecule is suspended in a highly ionized solution. But even this smaller form was larger than any known single polypeptide chain, so it seemed likely that RNA polymerase was constructed from a number of smaller polypeptides. No deep insights, however, came from electron microscopy done at MIT. It revealed approximately spherical particles 120 Å in diameter, compatible with a molecular weight of almost a million atomic mass units. Just before leaving for Paris to become a postdoc at the Institut Pasteur, John used phage T7 templates to demonstrate the synthesis of RNA products containing up to ten thousand bases. Conceivably their long lengths reflected failure of the purified enzyme to recognize normal T7 stop signals for RNA synthesis.

After John left for France, Dick Burgess, who had just arrived from Caltech, took up the challenge presented by RNA polymerase's large size. Every week Wally's technician Chris Weiss separated the proteins from thirty liters of *E. coli* cells into several fractions. The one containing the lactose repressor went to Wally and Benno, while Dick took the fraction containing RNA polymerase. The latter still contained many, many other proteins, requiring him to work out a glycerol gradient centrifugation procedure that yielded a highly active RNA polymerase preparation that he called GG. Starting late in 1967, he added a phosphocellulose column step to his recipe, leading to even purer RNA polymerase, called PC RNA polymerase.

In April 1968, Dick gave a talk at the Federation Society meeting in Atlantic City revealing that PC RNA polymerase particles were made up of small subunits, called α, and large subunits, called β. Two months later, at the Gordon Conference on Nucleic Acids in New Hampshire, he gave out details of his GG and PC purification methods, later sending written protocols to some ten interested labs, including that of Ekke Bautz at Rutgers. By then Wally's graduate student Jeff Roberts was also studying RNA polymerase. Initially using Dick's GG purified enzyme, he found it very active in transcribing λ DNA. But later, using Dick's even more purified PC preps, he inexplicably found no transcription of λ DNA.

Jeff's unexpected negative result made Dick wonder whether his PC purification procedure had removed an RNA polymerase component necessary for transcription of phage DNA. To test this possibility, he passed a fresh batch of GG enzyme over a PC column to see whether this again led to a loss of phage DNA's transcription ability. After getting the same negative result as Jeff, he showed that the PC enzyme could be activated by adding back a protein that had been extracted in purification. Experiments done the following week by our English postdoc Andrew Travers demonstrated that a single polypeptide chain, soon to be called σ factor, was the GG-enzyme-containing ingredient lacking in PC preps.

Less than a week passed before Dick heard that Ekke Bautz at Rutgers and his student John Dunn had obtained the same results using GG and PC enzymes following the protocol he had sent them in July. At my suggestion, Dick immediately contacted Bautz to see whether they would like their work included in a joint lab paper on the σ factor

At the 1970 Cold Spring Harbor symposium: Jeff Roberts with
Ann Burgess (left) and John Richardson with Dick Burgess (right)

to be submitted to *Nature*. Liking the idea, they came up to Harvard on November 6. For me their presence was also a distraction from the unbearable cliffhanger of the presidential elections that day. Richard Nixon's victory over Hubert Humphrey would become clear only late that night. Even the discovery of σ factor did not alleviate the gloom I felt the rest of the week at the prospect of a Nixon presidency.

On Monday morning, November 11, a silver lining briefly presented itself. In that day's *Crimson* I read the supersize headline "Pusey to Quit Harvard." The accompanying article reported that President-elect Nixon had paid an unannounced visit to Harvard, coming secretively at 9:00 P.M. the night before, to ask Pusey to become his Postmaster General. After a frank and comradely conversation, during which tea and vanilla wafers were served, Pusey accepted later declared to be "the greatest moment of my life." All too soon, however, the happy glow permeating my being dissipated. Lower on the page was another story reporting that the proposed JFK Library was to be moved from Cambridge to Bayonne, New Jersey, making me realize I was reading one of the hoax issues *Crimson* editors occasionally created, perhaps to assure themselves that no want of wit had consigned them to the *Crimson* rather than the *Lampoon*. After this monumental letdown nothing in the *Crimson* ever seemed funny.

When the manuscript announcing σ factor was submitted to *Nature* on December 2, it contained only data obtained at Harvard. The equivalent experiments from Rutgers had been done fewer times and were less complete. "Factor Stimulating Transcription by RNA Polymerase," by Burgess, Travers, Dunn, and Bautz, speedily appeared as a full article in January 1969. Before publication, I announced our lab's breakthrough in a lecture to Arthur Kornberg's perennially self-congratulatory Biochemistry Department at Stanford. It was a moment I never dreamed would come: our Harvard Biolabs demonstrating the biochemical competence to take on and best mighty Stanford.

In mid-December, Dick and Andrew had heard from Ekke Bautz and John Dunn that σ molecules disappear from cells infected with T4 phage, explaining why host bacterial RNA synthesis stops soon after phage infection. Phage DNA molecules likely coded for phage-specific σ factors that directed core RNA polymerase molecules to specific signals on their respective DNA. Over the next six months, Andrew Travers confirmed this conjecture and sent a paper to *Nature* in August called "Bacteriophage Sigma Factors for RNA Polymerase." In its conclusion, he speculated that σ-like transcription factors might be responsible for the massive shifts in RNA transcription patterns underlying the development of higher organisms. Sensing again a major paper on their hands, *Nature* editors rushed it into print in just over a month.

By early February, Dick had defended his thesis and started a Helen Hay Whitney postdoctoral fellowship that would let him remain at Harvard until his intelligent, blond wife, Ann, finished her Ph.D. thesis. She hailed from a prosperous, hardworking Wisconsin family and did her experiments on bacteriophage ϕX174 down the hall in Dave

Denhardt's lab. By then, Dick knew that not one but two β chains existed in the core (PC) RNA polymerase particle whose structure was $\alpha_2\beta\beta^1$. He and Andrew, furthermore, had preliminary evidence pointing to how σ functioned in the initiation of RNA synthesis. Its role was to direct the core $\alpha_2\beta\beta^1$ complex to appropriate starting sites to the DNA for RNA synthesis. RNA polymerase's enzymatic property is solely owing to its $\alpha_2\beta\beta^1$ core component.

The relative importance in controlling gene regulation of repressors and operators versus σ factor shifts still remained open. Bernie Davis at Harvard Medical School bet me a case of wine that my group would not discover a second bacterial σ factor over the next two years. Andrew then had let Bernie know of his tentative evidence for a σ factor controlling ribosomal RNA synthesis. Here time was not on his side, costing me a case of cabernet. But Richard Losick, a newly appointed junior fellow, had begun experiments in the Biolabs on how *Bacillus subtilis* forms spores, soon getting hints of a possible σ factor specific to bacterial sporulation.

The first visible recognition of σ factor's importance came in November 1969 at a meeting in Florence, Italy, on RNA polymerase and transcription. Dick Burgess gave the opening talk, coming from Geneva, where he was now a postdoc in Alfred Tissières's lab. Sponsoring the meeting was Lepetit, the Milan drug company whose rifamycin and rifampicin antibiotics had been shown to inhibit bacteria through binding to the β subunit of RNA polymerase. At this meeting, Jeff Roberts announced his recent discovery of a protein called ρ that halts RNA synthesis at specific stop signals on DNA molecules. Just before coming to Italy, he sent off his manuscript "Termination Factors for RNA Polymerase" to *Nature*, which published it in December 1969.

Though the low sun was not optimal for picture viewing in the Uffizi Gallery, the Palazzo Vecchio provided a reception site equal in its satisfactions to the science of the three-day gathering. Even being forced to listen to a minister up from Rome could not diminish the pleasure of knowing that the Biolabs were still at the center of how genes are regulated.

Remembered Lessons

1. Two obsessions are one too many

Experiments, like many speculative enterprises, are likely to require at least five times more effort than you initially guess. Being a really good anything—be it university president, violinist, securities lawyer, or a scientist—requires a virtually obsessive devotion to one's objectives. Dividing one's attention will give the edge to competitors who have the same talent but greater focus. For this reason, highly successful bankers who also claim to be accomplished cellists are often neither. Their banking reputation likely rests on the labors of talented associates working day and night, and their cello playing as likely suffers from the time lost to even the pretense of being a banker.

2. Don't take up golf

The moment golf clubs are first spotted in your trunk, you will be subject to constant ribbing. Only the rare few content to play occasionally with no fantasy of breaking 90 should even consider hitting the links. Once you become obsessive about bettering your personal best—say, now 94—your weekend science experiments cease. You have become a thank-God-it's-Friday scientist, always fighting not to fall too embarrassingly behind those peers who have sensibly chosen the less Zen but more aerobic thrill of hitting tennis balls.

3. Races within the same building bring on heartburn

A serious competitor aiming for the same objective inherently creates anxiety. It is emotionally draining to wish ill upon anyone constantly nearby, yet few, if any, human psyches are capable of turning off the visceral wish for an opponent to stumble. In science, the journey is not the destination; the destination is the destination. And so it is better that one's competitors be in a different city, if not country. Having them in the same building is a small model of hell, and not even an efficient one. Once Jacques Monod had two research teams working in adjacent labs attempt to make β-galactosidase in the test tube. Maybe he knew their respective leaders already disliked each other. Anyway, in that spring 1962 race, no one crossed the finish line.

4. Close competitors should publish simultaneously

Science works better when the winners do not take all. The agony of losing a very close race may break the spirit of a competitor who may again bring out the best in you. And so when you beat someone across the line by only a nose, offer to publish at the same time, if not back to back, in the same journal. Those scorekeepers in the know usually are aware what happened and will think more highly of you. Doing unto others may also yield reciprocal benefits the next time you are oh-so-close.

5. Share valuable research tools

Do not hog powerful new research tools or reagents. If Dick Burgess had not shared his GG and PC RNA polymerase preps with his lab neighbor Jeff Roberts, he unlikely would have been the first to discover σ factor. One hand washes the other. Though it was sharing his RNA polymerase protocols with Ekke Bautz and John Dunn that brought them into his game, they later immediately shared their discovery that σ factor disappears following phage T4 infection. This openness further helped Andrew Travers get an early start in the hunt for phage-specific σ factors.

IN THE FALL of 1967, Harvard gave me permission to become the director of the Cold Spring Harbor Laboratory while remaining a full-time member of the faculty. They did so upon realizing that Cold Spring Harbor's precious research and educational resources were on the brink of disaster and would likely disappear unless someone stepped in to make this unique Long Island institution financially viable. I was already a member of its board of trustees and knew its shaky state through my close friendship with its sharp-minded director, John Cairns. Born and bred in North Oxford, John exuded an ironic intelligence as well as an inability to seek help from individuals who had more power than warranted by the agility of their brains.

Since arriving from Australia in July 1963, John increasingly had come to see the biochemist Ed Tatum, chairman of his Board of Trustees, as a personal nemesis. On paper he seemed an asset. Tatum, then a professor at Rockefeller, had done research at Stanford in the 1940s on gene-protein relationships that led to his sharing the 1958 Nobel Prize in Physiology or Medicine with the Nebraska-born geneticist George Beadle. But Tatum was a polite plodder who would have gone nowhere but for Beadle, and later at Yale he was propped up by his graduate student and protégé Joshua Lederberg. The sharp intellectual crossfire of the geneticists and molecular biologists who spent summers at Cold Spring Harbor was not Tatum's cup of tea. Prior to his becoming chairman, he had attended only one summer symposium. Much to John Cairns's annoyance, Tatum arranged for all

*John Cairns (center) at the
1968 symposium*

trustee gatherings to be held in New York City at Rockefeller University. Avoiding the thirty-mile trip east, the chairman spared himself and the other trustees having to see the decrepit state of the some twenty-five buildings on the Lab's almost one-hundred-acre campus. At board meetings, Tatum's demeanor reminded me of that of Nathan Pusey. Neither knew how to handle individuals who presented unwanted facts.

By the time I joined the board, John operated in day-to-day uncertainty, even though he had converted the lab's $50,000 negative cash balance to a surplus of some $100,000. Three years of hard physical and mental effort had been required, including his oft cutting the Lab's many green lawns himself. At the same time, the Lab's endowment remained effectively zero, and its survival depended upon the success of its small number of scientists each obtaining one to several significant research grants, which supported not only their science but also the budget for administration, facilities maintenance, and the like. Apart from this grant income, the only other major barriers to insolvency were a few corporate sponsorships and the ever-increasing sales of the lab's annual symposium report, a must-have volume for anyone in molecular biology.

By mid-1966, John began to talk about quitting, and such talk only increased as Tatum showed himself entirely unperturbed by it. But several key junior scientists took note and in turn began to seek jobs elsewhere. In early 1967, John submitted his resignation, effective upon the selection of a successor. By then it was cold comfort to John that Ed Tatum would be gone even sooner. Replacing him as chairman was an H. J. Muller–trained geneticist from Texas, Bentley Glass, whose association with the lab went back to the late 1930s. Recently he had moved from Johns Hopkins to become provost of the new Long Island campus of the State University of New York, at Stony Brook.

Bentley knew that he alone could not dramatically improve the Lab's chances for survival. Though he was connected to innumerable government funding sources, no new monies would flow to Cold Spring Harbor until an appropriate new director was found. When I flew to New York City for the Lab's late October trustee meeting, I feared they might choose the German phage geneticist Carsten Bresch, then looking for an escape from an insecure job in Dallas. If Bresch were to come, he would continue the Lab's historically strong focus on molecular genetics. But I feared that he would see the job as a way station for him until a permanent, well-financed position was created in Germany.

Still, this was not an objection likely to sway my fellow trustees. Better a temporary leader than none, the others would counter. So I threw my hat into the ring, saying I would accept the directorship if I could simultaneously remain at Harvard as a full-time member of the faculty. In this way, the Lab trustees would be spared the need to find a source from which to pay the new director. Since his arrival, a five-year grant from the Rockefeller Foundation had covered John Cairns's $15,000 annual salary. But one could not count on an extension of this grant. As soon as I raised the possibility of my leading Cold Spring Harbor, further discussion about Bresch subsided. I sensed that if Harvard said yes, the job was mine.

Before coming down to New York, I had not consulted with anyone about the likelihood of Harvard's allowing me to hold two academic positions simultaneously. So immediately upon returning to Cambridge, I contacted Paul Doty, who had assumed his role as first chairman of the Department of Biochemistry and Molecular Biology. After receiving positive concurrence from its members, he wrote to Franklin Ford on November 22 proposing that I serve as director of the Cold Spring Harbor Laboratory for a five-year interval. During this time, I would continue my current Harvard teaching and committee responsibilities while being in Cold Spring Harbor for three days every two weeks on average. Only five days passed before Ford wrote with his approval, noting that he would have to pass my request on to the Harvard Corporation for official endorsement. I so wrote Bentley, who in turn asked other board members for a formal endorsement of my selection as director. On February 1, 1968, my new responsibilities began.

My decision to take on Cold Spring Harbor's troubles was in no small part sentimental. When I was there I was home. To me it was science at its best, where finding deep truths mattered more than personal advancement. Never had I known anyone to pull rank there or to get above himself. I could not contemplate the thought of its demise. As director, moreover, I could test my 1958 hypothesis that the cancerous potential of DNA tumor viruses is owing to the presence in their genomes of genes encoding enzymes that turn on DNA synthesis. It was too good an idea not to have a high chance of being correct, but because of space and funding limitations it had no chance of being tested at Harvard.

The sad underutilization of lab space at Cold Spring Harbor could prove a blessing. Its scientific direction could be changed fast without the unpleasantness of alienating excellent scientists already there who were oriented in the old way. To move swiftly into molecular biology, MIT in the mid-1950s had effectively fired its entire Biology Department, a move that generated much bitterness. This would not be necessary at Cold Spring Harbor.

On Sunday, February 4, I made my first public appearance as director. The occasion was the annual meeting of the Long Island Biological Association, whose membership initially was drawn from the owners of the great estates that had once dominated much of the landscape of the North Shore. Though the twenty years since the war had seen many big estates subdivided, there still remained, within a several-mile radius around the lab, the opulent homes of many of Harvard's most loyal and generous Wall Street benefactors. Thus, I thought, my Harvard professorship could prove as relevant as my Nobel Prize in mobilizing the local gentry behind our new cancer research objectives. Equally important was the high esteem in which John Cairns and his family were held by the Cold Spring Harbor community. That afternoon I publicly announced my hope that John would remain as a lab scientist and continue to live in Airslie, the large wooden manor house built for Major William Jones in 1806. Long part of the Henry deForest estate just to the north of the lab, it became the director's home in 1942. Being single and planning to be on site at most six to eight days a month, I did not need its many rooms. Before the association meeting started, I requested the even older Osterhout Cottage as my own residence. Alfred and Jill Hershey had lived there for several years before building a largely glass-walled house on land west of the lab.

At that time, my father was avoiding the winter cold in an old-fashioned resort on the west coast of Florida below Sarasota. This was his fifth winter there, the first having followed the mild stroke in November 1963. Once the awful shock of my mother's sudden death in 1957 had passed, Dad's broad, warm smile helped him make new friends among kindred souls, who valued books and Rooseveltian ideals. In particular, he met several quiet intellectuals associated with the experimental New College, on the grounds of the once expansive Ringling estate outside Sarasota. Two years before, he'd been proud to attend a lecture I gave to its students. The college's focus on the great books of Western civilization reminded me fondly of my University of Chicago years. His last Florida visit, however, had gone less well, as Dad's long-dormant stomach ulcer again opened up. Fortunately, it soon healed, and he felt confident enough to spend several spring weeks on a cruise to the Mediterranean before passing much of the summer on Martha's Vineyard in Edgartown's quaint Harbor View Hotel.

Still, he was only in middling health when he left my sister's Washington home after Christmas to again take up residence in Florida the year I became director. The persistent bad cough he'd developed over the holidays wouldn't abate in the Southland's warmth. But his Sarasota physician reassured my sister, Betty, several times over the phone that Dad did not have a virulent pneumonia. He was otherwise in good spirits, particularly when two *Atlantic Monthly* issues serializing *The Double Helix* appeared without generating a firestorm of criticism. He was also proud that his New College friends got a kick out of seeing me on the *Merv Griffin Show*.

Then, without warning, my sister called late one afternoon to report that Dad's persistent cough was never to go away. It was caused by an inoperable lung cancer, and the prognosis was that Dad had but a few months left. The two packs of Camels that he had smoked every day since college had finally caught up with him. Betty had gotten the grim news while I was en route to Cambridge from New York City after the lunch marking the publication date of *The Double Helix*. I was at my office when she finally reached me that afternoon. Then with me was the very pretty Elizabeth Lewis, the Radcliffe junior who on many afternoons assisted my secretary, Libby Aldrich's sister-in-law, Susie. Liz's appearance in the Biolabs several times a week to file reprints or to help me assemble successive drafts of *The Double Helix* invariably made me feel good. Conversely, I always felt lonely when she retreated back into her student life.

When she first came to Harvard, Liz thought about majoring in math, a subject that she had much enjoyed as a student at the Lincoln School in Providence, where her father, Robert Vickery Lewis, of Welsh and Yankee antecedents, practiced medicine. After his college years at Brown, he studied medicine at the University of Pennsylvania, where he met his future wife, the nurse Edith Mae Belle Irey, of Scots-Irish and Pennsylvania Dutch heritage. Being at a small Quaker school in no way prepared Liz for the Harvard math concentration, and she switched to physical science as a possible route to medical school.

*At my cousin Alice's wedding to James Houston in 1967; I am
to the right of the bride, and next to me are Betty's husband, Bob
Myers; my sister, Betty; my father; and William Weldon Watson.*

Our first effective date was unplanned, she coming with me at the
last moment for an early pre-supper get-together at Carl and Anne
Cori's home off Brattle Street. Afterward we drove along the Charles
River to Boston, where we saw an English movie at the Exeter Theatre.
Her exams were finished, and she was about to depart for a summer
job in Montana at a resort ranch above Yellowstone Park. It had
seemed a long summer when in early August a brief note from her
made me realize just how keenly I had been anticipating her return to
my office in the fall. Just after she got back to Radcliffe, we ran into
each other on Brattle Street near Sage's Market, which coincidence
gave me my second chance to drive with her into Boston. After lunch
on Newbury Street, we went into Bonwit Teller, the elegant shop
spread over the several spacious floors of what had been a gracious
city mansion.

Over the fall months, she had increasingly continued to forgo
evening meals at Moors Hall to join my father and me for supper at
the Hotel Continental. Upon his return from Martha's Vineyard in
August, Dad had chosen to move into the hotel, leaving his apartment

at 10½ Appian Way. It would save him the trouble of shopping, preparing meals, and tidying up. At the time I did not let on to Dad how my affection for Liz had increased over the past eighteen months. I knew he would worry that at nineteen she was likely to reserve her true affection for someone much closer to her own age.

As soon as I put down Betty's call, I asked Liz to stay with me for supper at the Continental. I did not want to be alone. It would be our first dinner together by ourselves. Afterward she did not go back to her dorm, telling me she did not want me to be alone that evening any more than I did. The next afternoon she left my office early to go grocery shopping on Brattle Street, planning to cook dinner that evening on the antique stove in my Appian Way flat. She had brought schoolbooks to read after dinner in the unheated alcove off the main room. The next night, when again we went together to Carl and Anne Cori's home for dinner, Anne knew she no longer had to find single girls to sit next to me.

Early the following morning, I left Liz to fly to Sarasota to collect a now very apprehensive Dad and bring him by plane to my sister's home in Washington. In 1964, after resigning from the CIA, her husband, Bob Myers, founded *The Washingtonian* magazine with his University of Chicago roommate Laughlin Phillips. Bob was its first publisher and Laughlin the editor. Just recently, Bob had become publisher of the *New Republic*, but too late for Dad, long a faithful reader, to take anything but a brief pleasure from seeing his son-in-law help run his favorite magazine of liberal politics.

Upon my return to Cambridge, I found myself all too soon scheduled to leave Liz again for the annual American Cancer Society (ACS) get-together of scientists and science journalists, this time being held in La Jolla, to the north of San Diego. Out of this meeting, it was hoped, would emerge optimistic press coverage to kick-start the ACS annual fund drive. Several months before, I'd eagerly accepted the invitation to attend, believing the meeting would help me focus on how to start up tumor virus research at Cold Spring Harbor. As it was to be held just before Harvard's weeklong spring break, there was also the possibility of Liz's joining me there after the conference.

To appear as a couple on a trip without causing a scandal, however, it would be necessary for us to marry immediately after Liz's arrival in California. Happily, she had no qualms, instantly accepting my proposal that we effectively elope. We decided not to let anyone know except for her parents in Providence. In the end, the only other person at Harvard in on our plan was my secretary. She found out when Liz came in saying this would be her last day of work. Susie said that Dr. Watson would be much disappointed. Liz replied that, in fact, he wouldn't be disappointed at all.

Before flying west I called Sylvia Bailey, the English-born secretary of Jacob Bronowski, the English polymath hired by the Salk Institute upon Leo Szilard's death, for advice about how Liz and I could best get wed in California. To my surprise, she called back the next day, saying that it would be faster to arrange a church ceremony than a civil ceremony before a justice of the peace. If I gave her the go-ahead, she would contact her friend the Reverend Forshaw, whose Mission-style church was in the center of La Jolla. In turn, Liz went with her mother back to Bonwit Teller's, this time no longer just looking but ready to bring home several outfits appropriate to the occasion and the many photographs we would take to send to relatives and friends by way of announcement.

At the ACS science writers' gathering, I spoke at length to Bob Reinhold, a former *Crimson* editor, now writing about science for the *New York Times.* In the article he soon wrote about my plans to turn Cold Spring Harbor toward cancer research, he remarked on the nervous way I held my can of Coke, having no way of knowing that this was no garden-variety tic but the anxious anticipation of my wedding the next evening. My nervousness disappeared as soon as Liz came off the plane. Her smile would always make me feel good. Shortly, we drove north of La Jolla to get a marriage license that would permit us at 9:00 P.M. on March 28 to be wed in the La Jolla Congregational Church. That I was not a churchgoer was of no concern to Reverend Forshaw, whose library prominently displayed one of Bertrand Russell's thicker tomes.

Upon our return to the La Valencia Hotel, we had an early supper at its Whaler's Bar before going on to Jacob and Rita Bronowski's one-story glass house in La Jolla Farms near the Salk Institute. Its stylish

Liz and I on our wedding day, March 28, 1968, in La Jolla, California

ambience was much better suited to wedding photos after sunset than was the church. Afterward, Liz met my small circle of La Jolla friends, who came to the La Valencia for a surprise party without knowing its purpose. We spent our first night as a married couple in one of the rooms looking out on the Pacific Ocean.

The next morning we telephoned my sister to tell her and Dad the news and to let them know that we would stop off in Washington on our way back to Harvard. I went to find postcards to let friends such as Seymour Benzer and Paul Doty know that "a nineteen-year-old was now mine." After a leisurely lunch, we drove east to see the desert plants blooming around Borrego Springs. We spent the night at Casa del Zora before driving through the Anza Desert to the Imperial Valley and from there to the village of San Felipe, some seventy miles south of the Mexican border. There we spent two nights in a hotel catering to fishermen, taking care not to get sunburned while spending much of Sunday swimming in the already warm waters of the Gulf of California.

Unknown to us was Lyndon Johnson's sudden decision to make a major address to the nation that evening. Only after we were back on the U.S. side of the border, driving across southern Arizona, did we learn that the night before, Johnson had announced he would not seek reelection. We hoped he would get us out of the quagmire in Southeast Asia before his term ended, but it seemed a vain hope. Though Johnson then presented the Tet offensive as a big setback for the Viet Cong, he had to believe otherwise, else he would not be stepping down.

By nightfall we were outside Tucson. The next day we admired thousands of tall cacti on an early morning walk in the Saguaro National Park. Dropping off our rented Ford Mustang at the airport, we caught the plane for Washington. Spring was in full bloom, allowing everyone at Betty's house, next to Glover Archibald Park, to half ignore Dad's awful prognosis as Liz and I shared the details of our wedding and the days afterward. Unexpectedly on hand was a photographer sent by McGraw-Hill's new magazine *Scientific Research*. Its forte was fast-breaking stories about scientists as well as science itself. Word that I had just married was already about, and they wanted a picture of Liz and me. The resulting photos revealed Liz a photographer's dream, and we were to be seen together on the cover of the April 29 issue.

The next morning we drove north for four hours to Cold Spring Harbor to see our eventual home. From Washington, I had let John Cairns know of our impending day trip, and the Lab arranged a special welcome dinner prepared by Françoise Spahr. Her husband, Pierre-François, was over from Geneva for six months to work with my former Harvard student Ray Gesteland, whom John Cairns had recruited to the lab staff a year earlier. But by the time we gathered in the main room of the big Victorian house at the Lab's entryway, the news of our marriage was eclipsed by horrid events elsewhere. In Memphis, an unknown assassin had just killed Martin Luther King Jr. Before driving off to Providence the next afternoon, Liz and I were interviewed separately by a reporter from Long Island's leading news-

SCIENTIFIC RESEARCH

McGRAW-HILL'S NEWS MAGAZINE FOR SCIENTISTS • APRIL 29, 1968

Nobelist Watson
and bride:
Challenge at
Cold Spring Harbor

*Liz and I graced
the cover of*
Scientific Research
on April 29, 1968.

paper, *Newsday.* To her dismay, he wound up suggesting in the article about us that her Radcliffe education would effectively lead her to a life of darning my socks.

At Liz's home, I met her father, a physician, whom I discovered to be, like my father, a keen reader and skeptic. In this important way, Liz and I had similar upbringings. Though her parents sent her to Central Baptist Church, its chief attraction to them was the music—Providence's best. That evening, my eyes kept drifting to the TV set and its images of the widespread race riots in the wake of the King assassination. The assailant was still unknown.

The next morning, April 6, was my fortieth birthday, and were it

Scientist Begins His and Lab's New Life

By David Zinman

Cold Spring Harbor—Dr. James D. Watson, the Nobel Prize-winning biochemist who is 40 years old today, returned yesterday from a brief Mexican honeymoon with his 19-year-old bride and could not resist showing off his laboratory here.

Watson, a Harvard professor and author of the controversial best-selling scientific autobiography "The Double Helix," became director on Feb. 1 of the renowned but deteriorating Cold Spring Harbor Laboratory of Quantitative Biology. The small lab, which has been near bankruptcy for the past 10 years, has concentrated on genetics. But Watson plans to launch a five-year, $5,000,000 fund-raising campaign and turn the laboratory's emphasis toward basic cancer research.

Radcliffe Junior

"I don't pretend to understand all his work," said the former Elizabeth Lewis of Providence, R.I., a pretty, green-eyed Radcliffe junior. She spoke as she strolled through the lab's wooded 85 acres and by its 30 buildings, some dating back to the early 19th Century. "But I think we get along beautifully despite the difference in our ages. Jim is a youthful person. He's forward-thinking and he's boyish-looking. But he's so busy with his work he needs someone to take care of the little things that are important at home." Mrs. Watson, who had a part-time job in her husband's Harvard office, married him March 28 in La Jolla, Calif. He was addressing a cancer seminar there.

Watson, a slender man who constantly runs his hands through his thinning brown hair while he's talking, won the Nobel Prize for medicine in 1962 with two British scientists. The trio discovered the structure of DNA, the molecule that governs heredity.

In an interview yesterday, Watson said he planned to devote his summers to the lab. In the winter, he said, he would spend about two-thirds of his time at Harvard and one-third at Cold Spring Harbor. "The lab will be increasingly emphasizing research on animal cells and viruses, Watson said, "and I hope it will become a major center for basic cancer research."

While viruses have been found to cause cancer in animals, no human cancer has yet been traced to a virus. However, some investigators, including Watson, believe viruses cause some forms of cancer. "The problem is you just can't experiment with humans," he said. "But it is so likely I don't have any doubt . . . . The challenge is to find what transforms a cell from normal to cancerous."

The laboratory, founded in 1894 on a quiet North Shore cove, is famous for its summer symposiums and teaching programs that attract world-famous scientists. In addition, its year-round staff has made key contributions in the development of hybrid corn, in fruitfly genetics, and, most recently, in the genetics of bacterial viruses and bacterial resistance to antibiotics.

But the laboratory has had a hand-to-mouth existence on research grants. It has virtually no endowment to use to maintain its $3,000,000 plant or to pay attractive staff salaries. Consequently, Watson said, it has not been able to expand or maintain its buildings and grounds adequately.

Watson's rejuvenation plans call for a $2,000,000 cancer research building, a $400,000 library, tripling the present staff of 12 researchers with doctorates, and doubling its summer student enrollment by enlarging the dormitories.

GEORGE LANE NICHOLS MEMORIAL

Watson Shows His Bride the Laboratory Yesterday

The Newsday *article, April 6, 1968*

not for Liz by my side I would have been feeling sadly old. We pushed on toward Cambridge just before noon to give Liz time for a Saturday afternoon of housewares shopping in Harvard Square. Her first big purchase was an ironing board that I carried back from Dixon's Hardware. Later enriching our Appian Way flat was a second silver candlestick, a gift from the Society of Fellows to complement the one given to me upon my becoming a senior fellow. Another early purchase was a big cookbook by Julia Child, a local resident, which Liz bought upon the suggestion of the woman at the Radcliffe registrar's office who recorded the change of Liz's name from Lewis to Watson. Its recipes proved much more satisfying to Liz to master than those in her organic chemistry class. Inevitably my days as a beanpole were soon to end.

As soon as Liz had taken her final exams, we made the five-hour drive down to Cold Spring Harbor, where the Lab had rented a house on Shore Road to let Dad live with us over the summer. He had been hospitalized several times while with Betty but now was pain-free enough to move in and out of our rented four-door Dodge. The new car spared Liz having to master jump-starting my MG TF. The day after we arrived, another multicourse meal, this time featuring lobster américaine, was cooked by Françoise. While eating it, we were horrified to learn that Robert Kennedy had just been shot dead in Los Angeles. I had pinned my hopes on him to win the Democratic nomination for president. Not since World War II had daily life been so frequently overshadowed by such a string of woeful events.

Nancy and Brook Hopkins began coming over to our Shore Road home to keep Dad company. Nancy was down for the summer with more than a dozen graduate students from MIT and Harvard, all focused on phage λ, working together in the vacant lab space underneath that of Al Hershey. Also about were Max and Manny Delbrück, back for Max's fourth consecutive year of teaching a course in the Animal House on the photosensitivity of the mold *Phycomyces*. Instantly I sensed their approval of Liz, and relief that I no longer would suffer from chronic restlessness. Over July, Dad's condition worsened to require a twenty-four-hour home nurse. The chemotherapy he was receiving from the Lab's local doctor, Reese Alsop, was mainly palliative. By month's end, however, the pain proved too great to treat at home, and he was admitted to Huntington Hospital before being settled in a nearby nursing home. He would pass only a night there before pneumonia mercifully ended his agony.

Two days later, Betty joined me in Indianapolis and together we drove north some one hundred miles to the small town of Chesterton, near Lake Michigan, where our mother had been buried eleven years before. We would have met at the Chicago airport, but the Democratic convention was in progress and the city was full of antiwar protesters tangling with unsympathetic policemen. That day we wanted to think about Dad and Mother, not Vietnam. Meeting us the next morning at the cemetery were Mother's nearby Olvaney cousins, whom we'd seen much of in Michigan City before our university studies. After lunch we set off to see the little bungalow on Chicago's South Shore in which we had grown up. The drive later through Grant Park was eerily peaceful. The night before, Mayor Daley's police had violently dispersed protesters attempting to camp out in its open spaces, onto which looked the windows of rooms in the tall hotels where the convention delegates were staying.

Upon my return to Long Island, the three-week-long course on animal cells and viruses had started. There I first met the Liverpool-born, twenty-eight-year-old Joe Sambrook. He had flown east to lecture on pox viruses, the subject of his Ph.D. thesis at the Australian National University. Over the past two years at the Salk Institute, he had shown SV40 viral DNA integrated into the chromosomes of cancerous SV40-transformed cells. His work had been the heart of Dulbecco's recent June symposium talk, leading John Cairns to suggest I approach Joe about leading our DNA tumor virus effort. Quickly sensing Joe's high intelligence and ambition, I offered him a position starting the following summer. He quickly accepted and wrote up a big grant proposal to the National Cancer Institute (NCI), which would guarantee the Lab an infusion of $1.6 million over the proposed five years. Obtaining this money was virtually a foregone conclusion, as there then existed more cancer research money than good applicants to use it.

In fact, the only reviewer with any misgivings about the NCI grant was Harvard's Charlie Thomas. He wondered about the risk to humans from working with tumor viruses at the molecular level. Could exposure to the monkey virus SV40 cause cancer in humans? We replied that we would follow the same procedures used in the Salk lab of Renato Dulbecco, who'd apparently worked safely with SV40. Furthermore, we knew that fifteen years earlier, SV40 had been an inadvertent contaminant of early batches of the polio vaccine with which several million individuals had been vaccinated, with no elevated incidence of cancer.

Over the February 1969 Washington's Birthday weekend, we stayed at Redcote, the home of Edward Pulling, the newly elected president of the Long Island Biological Association. Though raised in Baltimore and educated at Princeton, Ed had been born in England and served as a British naval officer during World War I. Upon his recent retirement as founding headmaster of the Milbrook School, north of New York City, he and his wife, Lucy, moved to the estate she had inherited from her father, the J. P. Morgan banker Russell Leffingwell. When Liz and I had first visited their eighty acres of fields and woods the previous summer, Ed pointed out the hidden ditch called a "ha-ha." It kept Lucy's horses from coming too close to the patio where we had cocktails before supper. Then also present was the journalist turned canny investor Franz Schneider, almost eighty, and his wife, Betty, twenty-five years his junior. Earlier in life, Betty regularly flew their seaplane from the dock by their house to and from New York City. Later they had us meet Ferdinand Eberstadt, who owned a large estate on Lloyd Neck that he would soon give to the Fish and Wildlife Service as a nature preserve to prevent the building of a nuclear power plant on adjacent land.

On each such visit down from Harvard, we eagerly followed the building of the new house we were to occupy upon Liz's graduation from Radcliffe. Initially we had planned to renovate the 175-year-old Osterhout Cottage, next to Blackford Hall at the heart of campus. The necessary alterations would require that I put up $30,000, then the value of my shares of the fast-growing pharmaceutical company Syntex, acquired after meeting the company's founder, Carl Djerassi, a chemist and the inventor of the birth control pill. At a Cleveland gathering of the American Chemical Society in May 1960 we had each received a $1,000 award. Upon returning to Harvard, I invested my prize money in Syntex shares, later using another $1,000 from my salary to buy more. Soon after the renovation of Osterhout started, however, the local contractor told us the deterioration was beyond repair. He offered to build for the same sum a new house almost identical to the one that was meant to emerge after the extensive renovation. We would get higher ceilings and central air-conditioning to boot. Our architect, Harold Edelman, saw only value in a brand-new home, and never would we regret the decision.

Over the spring break, Liz and I escaped the nationwide student unrest over Vietnam by going to the Caribbean, where my world renown paid off. Some ten years before, the wealthy European industrialist Axel Faber had set up a foundation to benefit Nobel laureates by subsidizing stays at exclusive hotels and resorts. For $10 per night, I had stayed at Le Richemond and the Hôtel des Bergues in Geneva. Having found the Dorado Beach outside San Juan more geared to golf than swimming, we moved first to St. Thomas and from there to St. John's to stay at the Caneel Bay Resort, where we sipped peach daiquiris with Abby Rockefeller's brother David and his new wife, Sydney.

Liz's last two months at Radcliffe took a direction we never expected. Less than a week after our return, University Hall was occupied by some three hundred students protesting the war. Many were members of the radical group Students for a Democratic Society (SDS). The afternoon of Wednesday, April 9, red and black banners hung from a second-floor window after the building's occupants, largely administrators, including Franklin Ford, were roughed up by the students protesting their earlier unlawful expulsion. SDS had been threatening violence for some time, no doubt encouraged by the effects of similar student uprisings elsewhere. Intending their actions to stop a war, those occupying University Hall saw no reason for their conduct to be governed by codes observed in times of peace.

That afternoon they proclaimed their occupation would end only if the university acceded to several demands, chief among them the expulsion of ROTC from Harvard. In fact, the Faculty of Arts and Sciences two months before had voted to deny credit for ROTC courses and not to give academic appointments to the military officers teaching them. ROTC's presence, commonly supposed to be the root of the trouble, by itself never would have led to the occupation of University Hall. The unstoppable chain reaction began when Richard Nixon became president and Harvard lent him Henry Kissinger as national security advisor. The student protesters had reached the limit of their patience with Nathan Pusey, who had failed to address the moral quandary in which the university found itself, and who two years earlier had branded the campus's self-proclaimed student revolutionaries as "Walter Mittys of the left."

Despite repeated warnings to leave University Hall and Franklin Ford's closing off Harvard Yard to all except its freshmen inhabitants, the SDS-led students gave no sign of budging. Though a lightning police raid had been talked about earlier as the best way to deal with building takeovers, no one was prepared for what happened next. At five o'clock the following morning, four hundred blue-helmeted, shield-carrying Cambridge policemen entered Harvard Yard and with tear gas and clubs forcibly removed the students, then barely awake, many banded together arm in arm. After less than fifteen minutes of this mayhem, University Hall was cleared. Most of the students were herded into paddy wagons and carted off to the Cambridge city jail to be charged with trespassing.

The rest of the Harvard student body, until then largely unsympathetic to the SDS gang, instantly ignited with indignation against the administration and, in particular, President Pusey. Police brutality had made martyrs of the student protestors. Fifteen hundred students gathered that afternoon in Memorial Hall calling for a three-day boycott of classes. Even angrier crowds formed later in Soldiers Field across the Charles River. More than five hundred law students, from a school never before known for radicalism, voted for Pusey to resign. As I walked that day into the Faculty Club for lunch, Liz was among the students lining the path to its front entrance to protest the raid. I had never before seen her make a display of political opinions.

Though the police raid's impact upon students lasted only until commencement, schisms developed within the faculty that would last for years. A liberal caucus was formed soon after the event. The group believed that without Nathan Pusey as president, the whole ugly affair would not have happened. By reacting so insensitively to student concerns about Vietnam, he stood out as a naked proponent first of Lyndon Johnson's Vietnam policy and later of Richard Nixon's. In contrast, a conservative caucus of roughly the same size assigned all blame to the student activists. How the offending students would be disciplined was not initially clear. Just before commencement, the liberal caucus felt semivictorious when a broadly constituted committee, including several students, voted for the temporary expulsion of only ten students, those known to have manhandled administrators during the April 9 takeover. Those in the conservative caucus had wanted many more students held accountable and for the sanction to be severe, ideally permanent expulsion.

Exacerbating the spring tensions was an emerging political activism among many of Harvard's black students. Two months before, the Faculty of Arts and Sciences had voted to set up an undergraduate degree program in Afro-American studies. Emboldened by the chaos following the raid, the more militant black students demanded Harvard go further and create a separate department whose faculty they could help choose. Outside the April 22 faculty meeting held at the Loeb Drama Center to consider this matter, one black student stood holding a meat cleaver. Inside, to my subsequent regret, I joined the many liberal caucus members favoring student input in faculty choices. I then realized that letting Harvard science students help choose future science faculty would have been nonsense. Subconciously I must have believed this proposed Department of Afro-American studies would not long stand the test of time. Its offerings would not give black students the hard facts that would let them thrive in competition with students of other colors.

By then the war had directly affected the lives of Harvard's graduate students in science. No longer could they automatically defer military service. If their draft number went against them, they soon might be off to Vietnam. To avoid that potential fate, several first-year graduate students joined me in Cold Spring Harbor for the summer of '69. There they might get deferments on the basis of involvement in cancer research. Two students stayed over the next two years helping Joe Sambrook get his SV40 work off to a fast start. To further make the Lab a force in tumor viruses, Joe and Lionel Crawford from London organized a two-week workshop in August that attracted, among others, Arnie Levine, Chuck Sherr, and Alex van der Eb. They all later became leaders in tumor virus research. The workshop's core was a small tumor virus meeting with some eighty participants.

Later that month, Jacques Monod led a big contingent from the Institut Pasteur to our meeting on the lactose operon. From that event was to emerge the first Cold Spring Harbor monograph, an enduring volume of work with its chapters edited by Jon Beckwith and my former student David (Zip) Zipser. (Soon after, Zip left a new tenured position at Columbia for Cold Spring Harbor to set up a bacterial genetic group in space below Al Hershey's lab.) During the meeting, we used our new flat-bottomed boat to take Jacques out into the chop of Long Island Sound. He wanted me to speed up, but I wouldn't. Unbeknownst to the visitors, Liz was three months pregnant with our first child.

That summer, the Welsh-born Julian Davies was at the Lab doing

research on yeast protein synthesis. Julian was rejoicing in his recent move to the University of Wisconsin, where he no longer had to listen to his former Harvard Medical School department chairman Bernie Davis bemoaning his left-wing faculty members. One day during the summer, Julian passed around the Lab notices for a demonstration protesting the investiture in Caernarfon of Charles Windsor as Prince of Wales. His summer assistant, the daughter of our local village police chief, saw the flyers and told her father of the planned action.

Given the tenor of the times, Chief McKensie would take no chance of a student demonstration getting out of control. At the exact time Prince Charles was to be anointed, he appeared, halting the unauthorized march along Bungtown Road, Cold Spring Harbor Laboratory's main thoroughfare. To my relief, Chief McKensie had no comparably reliable intelligence concerning the marijuana freely available at many Lab gatherings that summer. Getting a whiff of it myself at Jones Lab's first summer party, I took care to forgo further such gatherings. It was better not to know of things I was now officially obliged to stop.

Even before Joe Sambrook's arrival, I knew the success of his tumor virus group would require constructing new lab facilities to supplement the space he was to have in the James Lab. Initially I thought $60,000 would cover the costs of an annex on its south side. So Liz and I flew down to Long Island from Cambridge to seek this sum from the pharmaceutical heir Carlton Palmer, whose grand Tudor-style home was on nearby Center Island. At the end of the Sunday lunch, hosted by our neighbor, the Lab trustee and lymphoma expert Bayard Clarkson, I was told that I would have to find fifty-nine more donors. The Lab still badly needed an angel, as I soon thereafter told a reporter for the *Long Islander,* the local Huntington weekly founded by Walt Whitman. To my delight, the resulting article generated a call to me from the former Pfizer executive John Davenport, who had a summer home near Lido Beach on the South Shore. While still with Pfizer, he had run an RNA tumor virus effort and liked what I proposed to do at Cold Spring Harbor. Soon he and Ed Pulling joined Liz and me for a home-cooked lunch at Osterhout, after which I gave John a tour of the science then going on at Cold Spring Harbor. Less than a week passed before Davenport took me to lunch at the University Club in New

York City, where he told me that he was transferring $100,000 worth of Pfizer shares to the Lab. Though our new building's costs had risen to $200,000, I had by then raised the rest. Construction would start as soon as the winter snows melted.

Almost all the next academic year, Liz and I lived in Cold Spring Harbor, much enjoying views of the inner harbor through the big eastern-facing windows of our new white cedar house. After flying to Florence for the November 1969 meeting on RNA polymerase, we drove to Venice and from there traveled by train to a big resort hotel on Lake Constance. There a meeting was held to further advance Leo Szilard's scheme for a European molecular biology research and teaching institution. From nearby Zurich we flew to London, going up to Cambridge so Francis and Liz could meet. *The Double Helix* no longer upset him since he realized that it enhanced, not diminished, his reputation. Back in Cold Spring Harbor for the holidays, we eagerly awaited the arrival of Rufus Robert Watson in late February 1970.

In May, Nathan Pusey announced that after eighteen years as Harvard's president he would be stepping down in June of the following year. No one was surprised. His handling of the occupation of University Hall, while perfectly legal, had been hugely unwise, dividing the faculty into two angry camps that had no hope of being reconciled unless he went away. He would, in due course, move to New York City to become president of the Andrew W. Mellon Foundation. After raising piles of cash that let Harvard expand in many directions, he was now to have the far easier task of giving it away. Paul Doty and I could not help but note that his only obvious failure at fund-raising was in not finding a major donor for the proposed building to house his new Department of Biochemistry and Molecular Biology.

In the summer of 1970, for the first time in its history Cold Spring Harbor did not play host to visitors conducting research of their own choosing. No free space remained to house them or let them carry on meaningful research. Instead we used our resources to start a new course on the molecular biology and genetics of yeast. It would allow scientists interested in the essence of eukaryotic cells to learn the powers of yeast genetics. In the fall, we used a $30,000 donation from Manny Delbrück to begin renovating the derelict wooden Wawepex lab into a dormitory for sixteen summer students. Its beds would let us start a summer neurobiology course program. A new grant of almost $500,000 from the Sloan Foundation was used in part to cover the costs of renovating the top floor of the 1912 Animal House into lab and lecture space for neurobiology.

With the completion of the James Annex in January 1971, our tumor virus group came into its own. No longer did we have to play second fiddle to equivalent endeavors at the Salk Institute. Working now with Joe Sambrook were the Caltech-trained Phil Sharp and Ulf Pettersson from Uppsala, who brought to us adenovirus tumor research. Both came as super postdocs and soon became bona fide members of the James Lab staff. To support their independent activities as well as new cancer research efforts in space to be vacated by Al Hershey's impending retirement, Joe wrote up a big grant proposal to expand our NCI funding to $1 million per year. Later, site visitors gave it top priority, allowing our Cancer Center to come into existence as of January 1, 1972.

Even larger sums for research would soon become available. Richard Nixon had just enthusiastically signed the National Cancer Act in response to recommendations by the philanthropist Mary Lasker's Citizens' Committee for the Conquest of Cancer. Boldly it was proclaimed that if we can send men to the moon, we should be able to cure cancer. Though I did not believe there was sound logic in relating the two feats, I felt strongly that bigger infusions of federal monies were needed to let tumor virus systems point us toward the mutant genes that were known to cause cancer. So I spoke before the Citizens' Cancer Committee at one of its first meetings in the summer of 1970.

Under the "Conquest of Cancer" legislation, the president appointed not only the NCI director but also members of a new National Cancer Advisory Board, among which I was included for a two-year term to begin in March 1972. Before that, I had been advising NCI as a member of its Cancer Center Study Section. Directing the NCI then was the bureaucrat Carl Merrill, mindlessly generating a complex flow chart for the cure of cancer modeled after Admiral Rickover's successful Polaris missile program. The one such planning session that I attended in Washington accomplished nothing. I was hoping my

Ulf Pettersson in the lab at Cold Spring Harbor, December 1971

Joe Sambrook at CSHL in 1973

Jane Flint, Terri Grodzicker, Phil Sharp, and Joe Sambrook at CSHL in 1973

forthcoming National Cancer Advisory Board meetings would do more good. The several other pure scientists on the board also appreciated nonsense for what it was. Its clinical oncologists, however, wanted to create more comprehensive cancer centers. But such centers, already operating in New York City, Buffalo, and Houston, had shown no capacity to cure most adult cancers; I saw no reason for more of them other than to create more good-looking places to die. Only inspired science, not public relations, could lead to the knowledge that might let us finally cure most cancers. A real role lay ahead for our lab.

Remembered Lessons

1. Accept leadership challenges before your academic career peaks

By forty, I was the right age to begin directing a major research institution. My ever-increasing focus on cancer was best fulfilled by presiding over seasoned professionals, as opposed to graduate students and beginning postdocs. So I consciously made the decision not to run a research group at Cold Spring Harbor. This allowed me to focus my efforts first on recruiting scientists who cared as much as I did about the biology of cancer, then on finding the funding they needed to make their ideas work.

2. Run a benevolent dictatorship

I was appointed to provide a vision for Cold Spring Harbor's future, not to persuade existing staff members to share it. A research institution cannot ultimately be a democracy. Still, a wise autocrat takes the trouble to sell his program. I never made appointments without seeking informal counsel among relevant scientists already at Cold Spring Harbor. And in time, most new appointments arose from their sponsorship by scientists already there. To my knowledge, I never appointed anyone not wanted by others on our staff.

3. Manage your scientists like a baseball team

Sports and top-quality research have much in common. The best stars of each are young, not middle-aged, though occasionally someone older than forty can still be a formidable force either across the net or at the blackboard. No one can long remain a science manager unless constantly on the prowl for talented rookies able to move the game to the next level. Research institutions that let the average age of their staff creep up inevitably become dull places. Lowering your average age by constructing new buildings to create more space, however, is not the way to go. If the older buildings are not exuding vitality, they become mere financial drains. Equally important, good managers see the need to retire scientists who are no longer hitting home runs. Only individuals who continually reinvent themselves through new ways of thinking should enter middle age still part of your staff.

4. Don't make midseason trades

Once you have hired a scientist, give him or her an honest chance to hit the ball over the fence. Grand slams usually come only after a player has put in three to five seasons. Treating your players according to who's got the hot hand from one week to the next only drives down overall self-confidence. Consistency and steadfastness are the way to get the best out of your players. In this regard, handling the science of others is radically different from doing your own. As a scientist, you can profit from frequent quick course changes until the right path becomes clear.

5. Only ask for advice that you will later accept

Don't second-guess advice you've sought from a colleague whom you highly respect. If a smart friend advises you to hire one of his or her best students or postdocs, consider it a lucky break and just do it. Conversely, never ask advice from someone with an outlook on science alien to your own; people they recommend will share their values, not yours.

6. Use your endowment to support science,
not for long-term salary support

Generally, a bare minimum of any research institution's funds should go to salaries for older scientists. Under normal funding circumstances, those who have remained successful invariably get adequate research grants to cover themselves and then some. Revenue generated by the endowment is best used to support scientists at the beginning of their careers, when their grant support usually is not adequate to their research needs. Science is not a welfare state: feeding the young until they can feed themselves serves the greater good, while failing to cull the herd does not.

7. Promote key scientists faster than they expect

By promoting your best performers rapidly, you necessarily reduce money that might be given to those whose work no one would miss. Be generous with those you value: a salary increase below the rate of inflation is the universal sign to start looking elsewhere.

8. Schedule as few appointments as possible

When a scientist pops into your office or calls to schedule a meeting, learn then and there what he or she wants. You can avoid wasting your time and theirs by immediately saying yes to their request for a piece of equipment or a salary slot they legitimately need to move forward with their jobs. Even if so acting means promising money you don't yet have, do it. Your purposes will be served if their worries exclusively concern getting their work done, not currying favor with you. Any scientist on the scene knew clearly where he or she stood with me; I backed everyone I wanted to stick around.

9. Don't be shy about showing displeasure

When someone working for you says something stupid or in other ways makes your blood boil, express your anger immediately. Don't go about silently seething, letting only your spouse know you are upset. That is bad for your health and really not fair to those whose behavior has offended you. They very likely already fear they are in deep shit, and would naturally want to respond to your criticism sooner rather than later. When things are left to fester, the festering takes on a life of its own, sowing unproductive distrust. The fault might even be yours. If so, apologize as fast as you flared up. Being an ass occasionally is forgivable; being unable to admit it is not.

10. Walk the grounds

Your staff will seldom come to your office to tell you of impending bad news. Only when the bough breaks do you learn the awful truth. To stay ahead, it's best to make walking the grounds a part of every day. This allows you to meet those lower on the lab's totem pole and acknowledge their existence with a smile or word of praise. Equally important is to pop in on labs in the evening or on weekends. Those populated only during weekday daylight hours are likely going nowhere, while scientists at their benches on Saturday afternoons are seldom there killing time.

15. MANNERS MAINTAINED WHEN RELUCTANTLY LEAVING HARVARD

DURING my initial years as director of Cold Spring Harbor Laboratory, I was never the Lab's primary fund-raiser. As president of the Long Island Biological Association (LIBA), Edward Pulling saw this as his role. Living scarcely more than a mile away, he frequently popped over to give potentially generous neighbors tours of the Lab. Early in 1972, Ed and his fellow LIBA directors committed themselves to raising $250,000 over the coming year. With it, the Lab would winterize Blackford Hall as our year-round dining hall, construct a second annex to James Lab for cell culture facilities, and buy a handsome Victorian house on Bungtown Road, on the way to the sand spit, for postdoc housing. Ed believed professional fund-raising help would be a waste of time and money. His acquaintances wanted only his reassurance that their money would be put to good use. My job was to be on hand when Ed had a hot prospect.

Ed was on to someone big when he saw the need to track down Liz and me on a brief late June holiday north of San Francisco. From a phone booth in Inverness on the way to Point Reyes, I confirmed that we would be there the next Wednesday when he brought Charles S. Robertson to tour the Lab. Charlie's wife had recently died, and he was looking for an institution to which to donate his estate, some ten minutes away on Banbury Lane in Lloyd Harbor. Excitedly Ed told me that some years before, the Robertsons had made the largest gift to date in the history of Princeton University: $35 million. Their wealth came from Marie's family, the Hartfords, whose one A&P grocery store in

With Edward Pulling at the annual meeting of the
Long Island Biological Association, December 1976

lower Manhattan spawned in just two generations some fourteen thousand more across the United States and Canada. At the start of World War II, they were the fifth wealthiest family in the United States. Charlie and Marie's gift to Princeton had been announced as anonymous, but when rumors arose that the CIA was behind the new Woodrow Wilson School of Public and International Affairs, they had to go public lest the gift be tainted.

Joining Charlie on his visit was his long-valued New York legal counsel, Eugene Goodwillie. After Marie's sudden death, Goodwillie told Charlie to quickly establish a legal residence in Florida. In this way, his estate would not be socked with punitive New York inheritance taxes. By the same logic, it was best for Charlie not to hang on to the Long Island estate, to which he felt such deep sentimental attachment. Its land had once belonged to his mother's Sammis family, and his marrying into great wealth allowed him to reclaim it. To keep the land forever intact, rather than subdivided into two-acre building plots, Charlie resolved to give it to a nonprofit institution. To be settled today was whether Cold Spring Harbor Laboratory was the worthiest beneficiary of the Banbury Lane property.

Charlie Robertson with me and Liz at CSH, September 1974

The visit started off well, with Charlie sensing the frenetic pace of the James Lab tumor virus effort as well as the boldness of the picture-window offices and seminar room of the James Annex. My grand office alone spoke of how the Lab was again on an upward course. Before lunch, I walked the sixty-seven-year-old Charlie and even older Eugene Goodwillie up to the Page Motel to introduce them to our living history in the persons of Max and Manny Delbrück. Delbrück had just arrived from Pasadena for a three-week *Phycomyces* workshop. Then, rather imprudently, I led my not-so-nimble guests down the steep one-hundred-foot path to Bungtown Road. Neither visitor stumbled, though they were only half smiling when we safely reached Osterhout Cottage. There they joined Ed Pulling for a lunch that Liz had taken great pains to prepare. She ate virtually nothing as she jumped up and down serving several courses that started with consommé Bellevue topped with horseradish-flavored whipped cream.

After lunch, Goodwillie drove back to New York City while I followed Charlie to the sixty-acre estate that sloped down to the eastern shore of Cold Spring Harbor. Before the war, much of it was still farmland, but that afternoon only several empty chicken coops spoke of that history. For the past thirty-five years the main house had been a large whitewashed brick Georgian structure built in 1936, and in it Charlie and Marie had raised their five children. Below the house, halfway down the sloping bluff, was a large saltwater swimming pond, which the children when young used to enter raucously, I was told, via a long steel slide. But the children were now grown and, furthermore, recipients of generous trusts, leaving Charlie free to dispose of his estate without detriment to them.

I sensed Charlie wanted us to do real science on his lands, and so to be straight with him I had to confess, nervously, that dividing our research facilities into two sites was not realistic. Instead I saw the best use of his land and buildings as a high-powered conference center similar to that of the CIBA Foundation on Portland Place in central London. Toward that end, the high-ceilinged, seven-bay garage could easily be transformed into a perfect meeting room for thirty to forty people. That night, Liz and I went to sleep not at all sure that the gift of the Robertson estate was what we now needed. Spending time to raise monies for conferences on Banbury Lane would divert us from raising funds to expand our cancer research programs.

To our relief, Charlie took less than a day to reach a decision that far exceeded our most optimistic hopes. Late the next morning, we learned that he had decided it made no sense to give his estate to an institution surviving hand to mouth. He would soon have Eugene Goodwillie draw up documents establishing an $8 million endowment to support research on the lab grounds. In return, we would accept the gift of his estate, which he would separately endow with $1.5 million. It should generate funds covering the annual operating costs of his estate, including a large annuity to be paid to Lloyd Harbor in lieu of taxes. The estate would come with a covenant to the Nature Conservancy, preventing any changes to the building and lands except for the building of a new residence to complement the visitors' rooms in the main house.

The remainder of the day we walked about in a virtual daze, half worrying that becoming rich would destroy the Lab's unique way of doing science. But we soon returned to our senses and accepted the

Robertson House in the 1970s

generous offer. Even with the forthcoming Robertson monies, we had more expenses than funds to cover them. Many key Lab buildings remained habitable only during the summer, and fixing them up for year-round use could easily occupy the rest of the decade.

Over the past six months, we had used accumulated profits from symposium book sales to winterize and totally renovate Cold Spring Harbor's original Firehouse. After buying it in 1930 for $50, the Lab had rented a barge to bring the building across the inner harbor to a site next to Davenport Lab, where it was subdivided into three apartments for summer use. Handling the badly needed renovation in 1972 was a local builder, Jack Richards, who had joined the Lab staff the year before to oversee construction of the James Lab Annex. Large picture windows, to rival those of Osterhout and the James Annex, were installed, creating views on the inner harbor from each of the three apartments, converting them from utilitarian to spectacular. Richard Roberts, the English chemist soon to move from Harvard, bringing the Lab expertise in nucleic acid chemistry, would occupy the top-floor apartment with his wife and two children. Below him would be Ulf Pettersson and his family, leaving behind a cramped apartment in a barn on Ridge Road. The basement flat would house Klaus Weber and his wife, Mary Osborn, soon to come down from Harvard to learn how to work on proteins of animal cells grown in culture.

Klaus Weber had risen rapidly at Harvard since joining me as a postdoc in the spring of 1965 to do protein chemistry on RNA phages. Recently promoted to full professor, he did not yet have the research facilities that normally go with the rank. All his previous research triumphs had come from using microbial systems, but he foresaw a bigger future for himself in moving on to animal cells and their related viruses. To learn how to grow and use them, he had just been granted a sabbatical leave for a year's work at Cold Spring Harbor. Going back to Harvard afterward would make sense only if they could provide space specifically outfitted for work with animal viruses. Toward that purpose, in the spring of 1972, I helped prepare a big application to the National Cancer Institute for funds to construct an extension to the Harvard Biological Laboratories. Mark Ptashne would potentially join Klaus in the new space. He too was keen to work on cancer-causing retroviruses since taking the tumor virus workshop the previous summer (1971). The idea was spreading: MIT was thinking of proposing to use "war on cancer" funds to create a similar facility by converting a former candy factory virtually adjacent to its main campus.

Next on the facilities agenda of Cold Spring Harbor was the winterization of Blackford Hall, courtesy of LIBA's amazingly fast $250,000 fund drive. Unfortunately, Jack Richards found it painful to work with Harold Buttrick, the well-connected New York architect that a LIBA supporter had chosen for the project. A direct descendant of Stanford White, Buttrick thought himself of higher caste than the builders there to take orders. After the project ended, I quietly made Ed Pulling aware that Buttrick and Jack had irreconcilable differences.

During mid-July, Liz and I were briefly in England so I could take part in a forthcoming hourlong BBC *Horizon* program on DNA. Our second son, Duncan, was only five months old and so I initially did not want to participate. But Francis, normally allergic to TV exposure, was keen on the project. So for several days we were filmed walking through key Cambridge colleges or standing next to the bar at the Eagle, the pub where twenty years before we'd regularly eaten lunch, and where Francis had first brazenly announced our having discovered the secret of life. An unusual ingredient of that interlude was the constant presence of the producer's girlfriend Eva. Several years before, she had been crowned Miss World, and she still retained global dimensions. Sadly, though, she may have paid dearly for a picture-perfect figure, regurgitating meals, as Liz accidentally observed her doing in the washroom of the restaurant one evening.

That summer, Cold Spring Harbor Laboratory published *Experiments in Molecular Genetics,* a much expanded version of material taught by Jeffrey Miller two years before in our annual bacterial genetics course. Its intellectual sparkle and visual elegance would likely lead to its wide adoption and thus real money for the Lab. It was a bargain—maybe too much of one: more than 450 pages for only $11.95. Like Jeffrey and all our other authors, I also then wrote gratis for the Lab. At roughly the same time, I was about to finish two long introductory chapters for the *Molecular Biology of Tumor Viruses.* When first conceived, it was to be a short book. But it steadily grew to more than 750 pages in thirteen chapters, written by twenty-two authors who included David Baltimore and Howard Temin, who were to share the 1975 Nobel Prize for their research on retroviruses. Also writing much of the book was Joe Sambrook, whose great talents as a scientist I found equaled by his ability to produce succinct, readable prose as well as edit the lesser sentences of others, myself included. He refused, however, to share credit as one of the editors. The book spine later showed only the name of my former Harvard postdoc, John Tooze, then in central London helping Michael Stoker run the Imperial Cancer Research Fund Laboratory bordering on Lincoln's Inn Fields.

The Lab's next book was the small volume *Biohazards,* drawn from the proceedings of a meeting held in January 1973 at the Asilomar Conference Center near Monterey, California. Its three days of discussions were organized to codify lab procedures appropriate for working with tumor viruses. No consensus emerged, however, among the one hundred attendees as to what precautions, if any, should be taken.

Helping to organize the meeting as well as to edit the book was our recently appointed staff member Bob Pollack, continuing research on SV40-transformed cells that he began at New York University Medical School. Originally a physicist, Bob had anxieties like those of Charlie Thomas about the safety of tumor virus research. I sometimes shared those worries, and as a precaution discontinued positive air pressure in our James virus labs. Positive air pressure was widely used in microbiology, a relatively higher pressure in the room preventing microbial contaminants in the outside air from entering. By the time of the meeting we had put these rooms under negative pressure, with their air venting through HEPA filters to keep viruses from escaping into the outside air.

Of increasing concern to me then was the possible doubling in size of the fifty-slip Whaler's Cove Marina on the eastern shore of the inner harbor. As it was, it posed no real aesthetic threat to the tranquility of our waterfront. But it had been purchased two years earlier by Arthur Knutson, principal owner of the larger marina operations in nearby Huntington Harbor, who shortly thereafter proposed to use the adjacent Captain White House to expand the local yacht club. This would turn Cold Spring Harbor into a very busy port indeed. Though neighbors had legally challenged Knutson's proposal, we were advised they were likely to lose. If our harbor was to be saved, the Lab somehow had to buy the marina.

Our able administrative director, Bill Udry, recruited Jerome Ambro, supervisor of the Town of Huntington, to help us. His intervention was critical since the town of Cold Spring Harbor did not exist as a legal entity—the eastern shore of the inner harbor was actually a part of Huntington. Bill and Jerry came to lunch at Osterhout, where we looked out on the marina while enjoying Liz's poached oysters. I was never privy to how Ambro subsequently persuaded Arthur Knutson to sell us the Whaler's Cove site. It was too bad, I thought, the Lab did not have means then also to buy the sea captain's handsome house.

The Board of Trustees approved the purchase early in June, just a week after the Robertson Research Fund formally came into existence. But their concordance was not as routine as expected. Arguing against the acquisition was our nearby neighbor Walter Page. Long an important Lab friend and a trustee during John Cairns's first years as director, Walter had left the board when the Morgan Bank sent him to London for several years to run their European operations. Upon his return, we asked him to rejoin the Board, but he begged off, citing his growing Morgan responsibilities. But when Charlie Robertson became our benefactor, Walter knew he had to come back to make sure our new riches would not be squandered. Spending $300,000 to buy a marina was not Walter's idea of a prudent first expenditure. I, however, believed that not to make the purchase was surely to waste perhaps our only chance to forever preserve the inner harbor's pristine state. An ideal setting for science is not a matter of purely utilitarian considerations. Sensing our fellow trustees swaying in Walter's direction, however, I threw a Hail Mary pass, threatening to resign if the marina was not purchased. After I made my announcement, I left the James seminar room and walked back to Osterhout Cottage, about a minute away.

There Liz was entertaining Marilyn Zinder, whose husband, Norton, was attending the Trustees meeting. I announced my abrupt move, and we nervously waited some forty-five minutes until Bentley Glass came in to say that the trustees had just voted to purchase Whaler's Cove. I walked back with him to the meeting, rather sheepish at having got my way by reason of blackmail. I would never again challenge Walter, against whom one public victory was one too many. As Cold Spring Harbor's most respected resident and an old friend of the Lab, he should have been informed well in advance of the meeting of how strongly I felt. Harvard duties that spring, however, had kept Liz and me largely in Cambridge. There we were comfortably ensconced in a Harvard-owned Kirkland Place house, less than three hundred feet from Paul Doty's much bigger mansard-roofed mansion. It had become our Cambridge residence in the fall of 1971, giving us plenty of space in which to prepare for the impending birth of our son Duncan early in 1972.

I no longer had John Cairns to help me bat around the pros and cons of impending Lab decisions. In early March, he returned to England to take the directorship of Mill Hill Laboratory of the Imperial Cancer Research Fund. The previous four years, his position at Cold Spring Harbor had been happily stabilized and paid for by an American Cancer Society professorship. Soon after getting it, John threw a big wrench into the science of Arthur Kornberg by finding an *E. coli* mutant able to live without his famous enzyme DNA polymerase. Its existence soon led to several successive searches for alternative DNA polymerases. Without John's genetic approach, the inherent complexity of DNA replication would have remained unknown much longer.

The Cairns family's departure opened up the possibility of my family occupying Airslie, the large rambling wooden structure that had housed the Lab's directors for almost thirty years. Built in 1806 for Major William Jones, Airslie had come into the Lab's possession upon the wartime dissolution of the late Henry deForest's large estate to the immediate north along Bungtown Road. Before Liz and the children and I moved in, however, we undertook a badly needed massive renovation. The Lab had previously never had funds for anything except the occasional fresh coat of paint and, once, a new roof. Cold winter winds blew through Airslie during all the years of Demerec's and Cairns's occupancy.

An initial plan drawn up by a New York architect seemed wrong at first sight. He would have given Airslie a formal Federal style appropriate for rich New England merchants. Somehow we had to find an imaginative designer to give the place its own character. Luckily I had just read that the celebrated Yale architect Charles Moore was doing a low-cost housing project in nearby Huntington Station. The year before, staying in the "honeymoon suite" of Sea Ranch, the resort he created above San Francisco, Liz and I had much admired the daring of its multiple sloping roofs. I arranged for Moore's next visit to Long Island to include a brief visit with us. A week later, Liz and I were observing up close his playful mind reimagining Airslie.

Luckily Moore's plan was within the Lab's fiscal reach. Less than a month passed before we saw his final scheme while he was up at Harvard lecturing to architecture students. We were immediately taken with the way he opened up the front of the house into a three-story hall, giving Airslie for the first time a large central staircase. On our next visit down to Cold Spring Harbor I shared the plan with the trustees, a little apprehensive imagining what they might make of the bold way Moore created large, open spaces from tight, smaller rooms. In particular, I worried what our new chairman and nearby neighbor, Bob Olney, would think. Happily, Bob approved, provided the local preservation society didn't object. On their subsequent visit, the society's president declared that Airslie lacked any design features worth preserving. His only concern was with saving some ancient panes of glass. Though it was rather impractical, we resolved to keep the small 150-year-old glass pieces that ran along both sides of the front door. And so early in the fall of 1973, when Airslie was no longer needed for overflow summer housing, the fourteen-month project started. The final cost of just under $200,000 seemed embarrassingly extravagant for housing the director. Once we moved in, we realized that Moore's unique design gave Liz and me a way of life usually enjoyed only by the very wealthy. But it occurred to me that one day this would help us lure my successor; contrary to stereotype, most scientists are far from indifferent to the finer things in life.

The following year, Robertson Research Fund money let us readapt Jones Laboratory into a year-round neurobiology facility. Charles Moore and his highly talented young coworker Bill Grover imaginatively placed the four specialized neurobiology modules as freestanding aluminum-covered boxes accented by boldly colored wooden strips. Charles Robertson and his new wife, Jane, came to its dedication ceremonies. Earlier in the summer, the Robertsons had invited all the speakers at the June symposium on the synapse for a late afternoon party. It marked our gracious host's last year in the home he had so lovingly occupied for almost forty years. By then he had given up plans to build a modest summer home next door to our proposed conference center, accepting that his and Jane's future would be mainly spent in Florida at his large waterfront estate in Delray Beach.

Even before Airslie was gutted and the electricity turned off, Liz and I wondered whether it soon might be our year-round home. How long Harvard would continue to let me be away so much was not yet settled. Living in two places, moreover, would become less practical once Rufus reached kindergarten age. Early in 1973–74 I informed Harvard that I might move out of Kirkland Place when my spring term responsibilities ended. Many of its furnishings would go to Airslie and others to the house we had just bought on Martha's Vineyard.

I had wanted a summer home on the Seven Gates Farm on Martha's Vineyard since becoming aware more than a decade before of its thousand-acre vastness, on which corn was still grown. I knew several of the farm's civilized summer denizens very well, including our Cold Spring Harbor neighbor Amyas Ames, former chairman of Lincoln Center, who had presided over LIBA in the last year of Milislav Demerec's directorship. Owning one of its only thirty houses, however, seemed beyond my means until late August, when a Vineyard Haven real estate agent took us to a simple early-nineteenth-century farmhouse just put up for sale by a man about to retire to low-tax New Hampshire. Were we to sell the house on Brown Street in Cambridge, bought using my Nobel Prize monies, we could just cover the purchase price. The idea became irresistible once we imagined ourselves basking in the shade of the two magnificent American elms overhanging the wide farmhouse lawn. Equally important, only five miles away was Ed and Lucy Pulling's West Chop beachfront summerhouse.

From 1971 to 1973, my main teaching responsibility was a yearlong introductory undergraduate course on biochemistry and molecular biology (Biochemistry 10). In preparing for a spring 1972 lecture on DNA replication, I saw the need to point out the commonly accepted $5' \rightarrow 3'$ DNA chain elongation mechanism led to incomplete double helices with single-stranded tails. Some molecular mechanism had to prevent DNA molecules growing ever shorter during each round of replication. Until then, no one understood why identical, redundant DNA sequences were found at the two ends of all linear phage DNA molecules. Suddenly I knew why. Redundant ends allowed right and left single-stranded DNA tails to hydrogen-bond to each other to form dimers. Further cycles of DNA replication would lead to ever longer phage DNA molecules. No longer mysterious to me was why replicating phage DNA molecules are many times the length of infecting

*Salvador Luria, Nancy Hopkins, and David
Baltimore at the MIT Cancer Center in 1973*

phage DNA molecules. Excited by my brainstorm, I told it to my Biochem 10 students and, afterward, wrote it up for an article in an October 1972 issue of *Nature.*

Soon my main concern at Harvard turned to making the Biological Laboratories another major site for tumor virus research; after coming strong into the molecular age, Harvard was risking again being behind the curve. In April 1973, however, the National Cancer Institute turned down Harvard's application for construction monies for animal cell facilities. The proposal's reviewers were not convinced that the building addition would be used in a way that well served NCI's mission. True enough, given my Cold Spring Harbor responsibilities, I would likely never directly oversee a tumor virus lab in Cambridge. Nor was it clear whether Mark Ptashne would abandon gene regulation in bacteria to work on retroviruses. And Klaus Weber might go back to Germany were he offered a high-level appointment. In contrast, MIT's application for NCI construction funds was approved without a hitch.

Actually, it was a shoo-in with David Baltimore and Salvador Luria as its main drivers. Soon they would conscript two Cold Spring Harbor initiates into tumor virus research: Nancy Hopkins, after two years in Bob Pollock's lab, and Phil Sharp, after three highly productive years in James Lab.

Moving much too slowly were the joint efforts of the Biology Department and BMB to recruit a tenure-level RNA retrovirologist. Though discussions began in the fall of 1972, letters seeking advice from eleven referees did not go out until six months later, in February. The referees were asked to compare as candidates Mike Bishop, Peter Duesberg, Howard Temin, Peter Vogt, and Robin Weiss. Temin's name was high on most lists, although with the caveat that he was not a consistently good lecturer. Two respondents advised that we also consider Harold Varmus. Mike Bishop was not on top of any list except for Bob Huebner's, who called him a highly intelligent biochemist as well as a lucid teacher.

On July 10, 1973, I went to University Hall to see the economist Henry Rosovsky, who had just replaced his fellow economist John Dunlop as dean of the Faculty of Arts and Sciences. Dunlop had hurriedly taken on the role during the April 1969 University Hall occupation when Franklin Ford suffered a mild stroke. I was then BMB's new chairman, a task I was to have only until February 1975, when Matt Meselson replaced me. During my brief tenure, I wanted to ensure that Harvard somehow acquired animal cell facilities equal to those MIT would have in eighteen months. That morning with Rosovsky, I stressed how important it was for him to ensure Klaus Weber's return to Harvard, a commitment Klaus could not make until up-to-date animal cell facilities were available in the Biological Laboratories or, better still, in the proposed new animal cell annex. Six weeks later, I took Klaus to Henry's offices to give Henry personal assurance that Klaus would come back to Cambridge if he could continue the experiments he and Mary Osborn had started at Cold Spring Harbor. By then Harvard decided to resubmit its application to NCI for construction monies, now opting for a completely separate new building on roughly the same site proposed for the annex. Several of the Biology

Department faculty had become queasy about living with possible biohazards in their midst.

Harvard's president, Derek Bok, became actively involved on November 11 when the ad hoc committee met to consider whether Howard Temin should be offered tenure. I appeared as an early witness, sensing there would be no objection. Soon the question became whether he would agree to come. To answer it, he and his wife came to visit Harvard in mid-December. Their main host was Matt Meselson, a close friend of Howard's since their days together at Caltech in the late 1950s. I was worried from the start that his population geneticist wife, Rayla, would not want to leave the University of Wisconsin at Madison. Matt, however, thought we had a good chance of getting them to join us.

In January 1974, Henry Rosovsky scheduled a day to hear everyone out about the proposed NCI grant resubmission. Carroll Williams expressed concern that receipt of NCI monies would force Harvard to use all of the government-financed space for work with cultured animal cells and their viruses. If the federal funds came forth, he thought BMB should relinquish some of its space in the Biolabs to let the Cellular and Developmental Biology subdepartment increase its numbers. In contrast, I argued for Harvard's quickly recruiting several animal cell hotshots who would create a high-powered cancer center like the one at MIT. It was my belief that all outstanding research on animal cells would easily fall within the purview of the NCI mission. Until we understood how cells sent out and received molecular signals to divide, we could not get at the essence of cancer, and understanding the mechanism was plenty to keep an animal cell group busy. I then told Henry the building would function best with a director reporting directly to the dean, which Carroll Williams resisted, seeing it would diminish the power of department chairmen. But I felt the chair's usual three-year term did not allow him or her to take on necessary long-term funding objectives.

The animal cell building proposal went to NCI in January just ahead of the deadline. The cover letter, signed by Derek Bok, was essentially written by me, with the requested construction money adjusted to $5.76 million, reflecting our architect's prediction of further inflation, which was then bedeviling the entire American economy. In its final form, the proposal was my vision—not Carroll Williams's—of how animal cell science should proceed at Harvard. But its realization much depended upon whether Howard Temin joined our faculty. More than a month of uncertainty passed before Howard finally turned us down, saying his wife saw her life diminished by moving to Harvard. No job she might find in Boston would be as good as the one she had with the population geneticist Jim Crow in Madison. To keep our NCI application alive, an offer was made to the English retrovirologist Robin Weiss. But soon he turned us down, as he had MIT the year before. My dominos continued to fall when Klaus Weber in early April formally accepted an offer to head a Max Planck Institute in Göttingen, Germany. It would provide even better facilities than could be fixed up for him within the confines of the Biolabs. Harvard then had no choice but to tell NCI it no longer could claim a future in tumor viruses.

Klaus's decision was already 90 percent made when Derek and Sissela Bok invited Liz and me for a late March Friday night dinner at Elmwood, the gracious old wooden house just off Fresh Pond Parkway where they lived with their four children. Upon Derek's becoming president, they chose not to live in the formal Quincy Street fishbowl successively occupied by Lowell, Conant, and Pusey. Two years younger than I, Derek had concluded three very successful years as dean of Harvard's Law School. A graduate of Stanford, he was the first president of Harvard not a product of the college. That evening I tried to forget about the animal cell biology fiasco and Harvard's lack of a tumor virus future. I realized there was no longer a good reason for Cold Spring Harbor and Harvard to remain closely connected. Derek graciously kept our conversation on other matters, knowing only too well that my heart was now mostly at Cold Spring Harbor. Harvard had no one leading it into the future in the way David Baltimore was blazing the way for MIT biology.

Two weeks later, I drove over to the glass-faced MIT biology building for a meeting hosted by Paul Berg. There, with David Baltimore's help, he assembled a small group to discuss implications of the powerful new recombinant DNA technology developed at Stanford. Phil

Handler, the president of the National Academy of Sciences, had asked Paul to come up with an appropriate response to a letter published in the September 21, 1973, issue of *Science*. The academy had been called on to offer guidelines for recombinant DNA experiments that might create biohazards not only for the lab worker but also for the general public. That morning our small group, which included Dan Nathans and Norton Zinder, concluded the matter would best be dealt with by a much larger group assembled at the same Asilomar, California, site where we had considered potential biohazards of tumor virus research the year before. Until Asilomar II could be held, likely early the next year, we proposed a worldwide moratorium on recombinant DNA experiments in a letter to the journals *Nature* and *Science*. I then visualized the Lab publishing Asilomar II's proceedings. Unlike our first biohazard book, this one I expected to make real money.

All the world's major tumor virologists assembled three months later for the Lab's annual early June symposium. Joe Sambrook organized the DNA tumor virus sessions, and David Baltimore put together the ones on RNA retroviruses. Between Renato Dulbecco's introductory talk and David's concluding summary, there were 116 presentations, out of which 101 manuscripts were generated. They would fill our first two-volume symposium proceedings, consisting of almost twelve hundred pages. Though no scientific bombshell exploded, the meeting's highly charged atmosphere made it likely that a deep truth would emerge at any moment. More presentations came from Cold Spring Harbor scientists than from the faculty of any other institution, even London's better-funded Imperial Cancer Research Labs in Lincoln's Inn Fields. Klaus Weber and Mary Osborn notably reported upon their purification of the SV40 antigen. In their talk they provided strong presumptive evidence that the T antigen was the product of SV40's A gene, one that functions early in the SV40 life cycle as well as in SV40-transformed (cancerous) cells. Further studies might soon convincingly show it to be the primary cancer-causing genetic unit on SV40's small circular chromosome.

I spent much of the remainder of the summer on Martha's Vineyard preparing the third edition of *The Molecular Biology of the Gene*.

Next to our old farmhouse was a small barn, whose large central room provided an ideal writing space. I was getting invaluable feedback from several science-oriented Harvard and Radcliffe students, who later extended the glossary and corrected the final proofs. Doing the many needed new illustrations was Keith Roberts, by then running his own plant cell biology lab at the John Innes Institute in Norwich, England. As a postdoc at Cambridge five years before, he had created the new drawings for the second edition as well.

Over the following academic year, I was again on leave, working full time at Cold Spring Harbor at a salary identical to what Harvard would have paid me for teaching. Our settled residence in Cold Spring Harbor allowed Liz to take two classes per week at the New York School of Interior Design. Often sitting near her was the petite, blond Barbara Lish, wife of the writer Gordon Lish, then America's most influential arbiter of fiction, whom we befriended. Most unexpectedly we bumped into the Lishes at an early December gathering of intellectuals on the Florida coast. Arthur Schlesinger, Gunnar Myrdal, Saul Bellow, Vernon Jordan, and I had all been assembled just north of Daytona Beach with the unexpressed purpose of drawing attention to ITT's big beachfront development called Palm Coast. It was still a day when public intellectuals could sell real estate. Attracting us to this most unlikely gathering was the generous $4,000 honorarium, a much more substantial monetary award than normally given for intellectual chitchat. Barbara and Gordon were there in pursuit of Truman Capote. At the meeting, Gordon persuaded Capote to let *Esquire,* where he reigned as "Captain Fiction," to serialize his newest opus, *Answered Prayers.* Before arriving at Palm Coast, we visited Disney World, where Duncan, just shy of his third birthday, screamed all through the jungle boat ride.

We were just a month settled into Airslie, its new picture windows alluringly draped with Swedish cloth we found in the D&D building on Third Avenue. Its many rooms let Liz invite her parents and her two brothers and sister, as well as her aunt from California and grandmother from Philadelphia, to spend Christmas day with us. But the big Christmas feast, preparation for which included many hours basting two geese, did not go as planned. By the time the fowl were on the dining table everyone except the schoolteacher aunt and physician dad had come down with twenty-four-hour retching flu. The night before, we had received all the families in Lab housing for warm Christmas grog. I didn't know whether they had brought the contagion or whether one of Liz's family members was its origin. Fortunately, there was no sign of a Boxing Day epidemic.

That year, the newly winterized Davenport Lab was utilized by three supermotivated yeast geneticists on sabbaticals: David Botstein from MIT, Gerry Fink from Cornell, and John Roth from the University of California at Berkeley.

After Christmas, our yeast trio and the tumor virologists began to discuss what should happen at Asilomar II, scheduled for February 1975. I increasingly worried about restrictions that might be imposed on the use of recombinant DNA technologies to clone putative cancer-causing genes. In fact, these procedures would greatly reduce whatever risks we were now incurring using live SV40 virus or adenovirus 2. Our call for a moratorium, however, created the mistaken impression, magnified by each successive press conference, that working with recombinant DNA was a potential major public health hazard possibly equal to nuclear weapons. Even before the meeting started, Joe Sambrook had been asked to join fellow tumor virologists in coming up with guidelines that could only retard the development of recombinant DNA technology.

When I arrived at Asilomar, I found that virtually all the 140 participants were inclined toward accepting restrictions of one sort or another. Only Stanley Cohen, Joshua Lederberg, and I thought they were the wrong way to go. To no avail we voiced the impossibility of regulating an unquantifiable risk. Harm to someone or something had to be demonstrated before regulation could be rational, and to our knowledge no tumor virologist had come down with a cancer likely to have been caused by lab exposure. But for Paul Berg and his Asilomar II co-organizers, there seemed no way out of accepting some form of NIH-imposed guidelines. If we attendees did not accept them, the wrath of public opinion would surely descend upon all of us. And if we did not propose them they would be imposed upon us in more draconian form. At the meeting's end, virtually all participants warily voted to approve the mildly restrictive rules prepared by the several working groups. If the public found them satisfactory, recombinant DNA experimentation should not be too badly set back. On the small feeder plane taking participants back to the San Francisco airport, however, I was full of foreboding. I believed that trying to look good, as opposed to doing good, could only backfire.

A week after Asilomar I flew up to Boston to speak at the dedication of MIT's Cancer Center. In my talk, I offered my view on how to fight the escalating "war on cancer." I proposed that money would be best spent initially on creating centers filled with Ph.D.'s, as opposed to M.D.'s. I did not see the big clinical cancer centers then as having the potential to attract the very bright young scientists who could find the molecular essences of cancer. And without those molecular keys all the money in the world would only little improve what clinicians could do. Only after my speech did I learn that an inexperienced young stringer for the *Washington Post* had been in the audience. To my horror, the next day the *Post* ran his story over the headline "Nobelist Calls War on Cancer a Failure." I immediately wrote Dick Rauscher, the RNA tumor virologist now heading the NCI, to say that I had been badly misquoted. Fortunately, someone on his staff, Phil Stansley, also heard my talk, and backed me up.

With my $1,000 MIT honorarium I soon acquired for the Lab a Milton Avery–like abstract painting by the talented Long Island artist Stan Brodsky. It gave real style to the fireplace room of Blackford Hall until it was damaged by a large spoon thrown during a summer banquet food fight. After repairs costing almost half the original purchase price, it went back on the same wall until the next food fight damaged it again. This time the harm was slight, and it was only a few days before its subtle red, pink, and blue colors could again be admired.

In the fall of 1975, I resumed teaching at Harvard, flying up to Boston to spend Sunday and Monday nights at the Harvard Faculty Club. My lectures on tumor virus and animal cells were updated versions of those I had given three years before, using as a text the Lab's monograph *The Molecular Biology of Tumor Viruses.* This was to be the last course I would teach at Harvard. Matt Meselson was unwilling to appeal to the dean for an exception to Harvard's long-standing prohibition against sharing faculty with other institutions. And so I was informed that, as of July 1, 1976, I would no longer be a professor at Harvard. It was a situation of my own making, but all the same I was much annoyed, if not insulted, since Jack Strominger had recently become director of research at the Dana Farber Cancer Institute across the river while retaining his professorship in our department. Jack, moreover, now was being paid by both institutions, while I would have been content with only one salary if I could keep both jobs. There were many things I knew I would miss about Harvard, but by far the first would be its students; the obligation to lecture to them forced me to extend my own thinking, and occasionally the most extraordinary ones came down for research at Cold Spring Harbor, enriching the intellectual fellowship there.

Just after Christmas, I flew to the West Coast with Liz, Rufus, and Duncan for a two-week visit that started in southern California, where for a week we stayed in an apartment on California Avenue just west of Caltech. There Max Delbrück had arranged for me to give a lecture honoring the recently deceased Jean-Jacques Weigle. I had always enjoyed Jean's nimble brain both at Caltech and when visiting him in his hometown of Geneva, where he did phage experiments in the summer. Afterward we drove up to San Francisco, where W. A. Benjamin held a book party to mark the appearance of the third edition of *The Molecular Biology of the Gene.* Like the first and second editions, its sales over five years would approach a hundred thousand copies.

Then working at Cold Spring Harbor developing a powerful new way to clone genes was the thirty-two-year-old Tom Maniatis. His highly innovative experiments at Harvard as a postdoc with Mark Ptashne had led to his recent appointment as assistant professor. Initially he was to come to Cold Spring Harbor Laboratory for only a year for work, returning to Harvard upon its building of facilities for animal cell experimentation. In collaboration with Argiris Efstratiadis of Fotis Kafatos's Harvard lab, Tom was using messenger RNA molecules

*With Rufus (right)
and Duncan in 1973*

from red blood reticulocytes as templates to make full-length double-stranded DNA copies of the β globin gene. No biohazard possibilities could arise from such experiments, so despite the recombinant DNA moratorium, Tom and his four young collaborators were able to move full steam ahead in their Demerec Laboratory space.

By then, his well-directed intelligence had also caught the attention of Caltech's biology department. Caltech let Tom know it was prepared to make him a tenured member of its faculty. Upon learning this, Mark Ptashne got BMB to ask Henry Rosovsky to move quickly in assembling an ad hoc committee and win approval to offer him a

May 27, 1976

Dean Henry Rosovsky
5 University Hall
Harvard University
Cambridge, Massachusetts   02138

Dear Henry:

Regretfully I now write to submit my resignation as Professor of Molecular Biology as of July 1 of this year.

As each year passes, I find myself more and more unable to function simultaneously as a Harvard Professor and as the Director of the Cold Spring Harbor Laboratory. And given the uniquely strong support to the Cold Spring Harbor Laboratory from its neighboring community, I feel I have no choice but to stay in Cold Spring Harbor to help direct the many research programs we have instituted on the nature of the cancer cell. Many of the recent large gifts to the Lab have been made with the anticipation that I shall continue as Director and to act differently would be a serious breach of faith.

Leaving Harvard will not be a simple event - I have far too many memories of the high quality of its faculty and students to easily forget the pleasures that can go with being a member of its faculty. Equally important has been the fact that I always felt that when the chips were down Harvard had no choice but to act wisely. Otherwise it would all too soon become like the rest.

I must also express my deepest appreciation to Harvard for having allowed me to function in this dual capacity for so many years. When I approached Franklin Ford some eight and a half years ago, the task of running Cold Spring Harbor Laboratory seemed easily manageable, and I naively thought that my functioning as a Professor would be unaffected. But then I was unmarried and free of any real responsibilities. Now, however, with two small sons and a wife, I seem to be always preoccupied, and I must not attempt to do more than I can do well.

I shall miss Harvard.

Yours sincerely,

J. D. Watson
Professor of Molecular Biology

JDW/mh
bcc: Dr. M. Meselson

Resignation . . .

tenured associate professorship. In February I went back to Harvard to attest to Tom's accomplishments before Derek Bok. As expected, Tom was approved.

The worry for Harvard all too soon became the possibility that Tom might choose to accept Caltech's offer anyway. The ever more vocal in-house opposition to recombinant DNA experimentation within the Biological Laboratories was hardly an inducement to prefer Harvard. The newly appointed assistant professor Ursula Goodenough and George Wald's wife, Ruth Hubbard, also a scientist, were claiming

HARVARD UNIVERSITY

FACULTY OF ARTS AND SCIENCES

OFFICE OF THE DEAN

5 UNIVERSITY HALL
CAMBRIDGE, MASSACHUSETTS 02138
(617) 495-1566

June 3, 1976

PERSONAL

Professor James D. Watson
Cold Spring Harbor Laboratory
Bungtown Road
Cold Spring Harbor, New York 11724

Dear Jim:

    A few days ago, when you handed me your letter of resignation, you caught me somewhat by surprise. I had heard from Derek and Matt a good many months ago that you had decided to remain full-time at Cold Spring Harbor, but you had not communicated directly with me. Since I requested to see you for entirely different reasons, I was not really prepared to deal properly with your letter and all that it implies.

    Your resignation is an event that I will always regret, and I regret particularly that it occurred during my term as Dean. I understand your reasons for leaving, but they do not make smaller the loss for your department or our Faculty. The fact that you were associated with Harvard for a long time will always reflect great credit on our scientific effort, and the University will remain indebted to you.

    I know that your colleagues will miss you very much. I feel the same way.

As ever,

Henry Rosovsky

HR/pm

. . . and response

recombinant DNA experimentation using *E. coli* might put the women in the Bio Labs at risk of cystitis. Both should have known that over the past ten years live *E. coli* cells had been regularly ground up by the pound by men and women alike without a single case of illness.

Initially Mark and Tom did not worry, knowing that Henry Rosovsky was not one to be intimidated by such nonsense, whose source was mainly left-wing elements that had been gaining ground on campus ever since the occupation of University Hall, and which after Vietnam had been further energized by Watergate. It did not matter to Rosovsky that at a public meeting in late May more students had been moved by the rhetoric of Richard Lewontin, Harvard's population geneticist, railing against future capitalistic exploitation of DNA research than had been stirred by my pleas to get on with cancer research using recombinant DNA. Henry Rosovsky sided with scientific progress and gave Harvard scientists the go-ahead to continue the controversial research. So Harvard's "science for the people" leaders took their case to the Cambridge City Hall and its populist mayor, Alfred Vellucci, always keen to put the Harvard elite in their place. At George Wald's urging, he and his fellow councilmen held hearings on June 27 and July 7, 1976, after which they voted for a three-month moratorium on recombinant DNA research within Cambridge city limits. Tom now felt going back to Harvard would be to enter a state of chaos, and so he accepted Caltech's offer, as feared.

Before reaching that decision, Tom had seen me very angered by Harvard for very different reasons when I appeared in his Cold Spring Harbor lab late in the evening directly following my return from several days in Cambridge. Derek Bok had invited me to his office in Massachusetts Hall, and I was expecting that Harvard would bid me farewell in some meaningful way. But for my presence at the Biological Laboratories for the past twenty years, its science would have commanded much less attention from the outside world. Wally Gilbert might very well still be a physicist, while Matt Meselson, and likely also Mark Ptashne, would be teaching in California. To my dismay, Derek's goodbye was entirely perfunctory, giving not a hint that my departure was any loss for Harvard or any detriment to its future.

At the start of June, I flew back to Boston for my last visit to University Hall as a member of its faculty. That afternoon, Henry Rosovsky and I tried hard not to focus on the fact that I was soon to be gone. It was easier to talk about George Wald's pretentious irresponsibility in opposing recombinant DNA. After twenty minutes of not wanting to say goodbye, I thanked Henry for trying so hard to get Harvard on the animal cell bandwagon. If I had not become irreversibly committed to Cold Spring Harbor, together we would have won. Realizing it was almost time for his next appointment, Henry, to my surprise, revealed that in looking over my salary history he noticed that I had always been paid too little. It was his way of saying he liked me. He and I knew that I would be sad walking out of Harvard Yard that day. Even I was not entirely immune to the old chestnut that there is no life after Harvard.

Remembered Lessons

1. Avoid boring people

Never make dull speeches that easily could be delivered by someone else. Predictable words naturally compel audiences to tune out and lock their pocketbooks. Just as tedious is bringing small groups of busy people together for committee meetings with no opportunity for them to offer real input. This is on a par with holding meetings where talk is not followed by meaningful decisions. In both circumstances, committee members will likely soon stop attending gatherings they know will be a waste of their time. Not boring others, of course, requires that you take pains not to become boring, as often happens when you begin to bore yourself. A leader's mind must continually be reconfigured through exposure to new patterns of acting and thinking. Reading the same papers and magazines as everyone else around you is not likely to make you an interesting dinner guest, let alone alter your consciousness. In my case, a subscription to the *Times Literary Supplement,* courtesy of my father-in-law, made me more interesting to sit beside than someone whose diet was limited to *Time, Newsweek,* or the *Economist*—or *Nature* for that matter.

2. Delegate as much authority as possible

Administrators, like scientists, do their jobs best when left alone to do them, freed of the irksome impression of simply carrying out someone else's will. My growing avoidance of micromanagement left me always available to give advice about matters that were not obvious. On learning how my staff wanted to proceed, I generally gave them a go-ahead. Few mistakes were made that closer oversight might have avoided.

3. Institutions are either moving forward or they are moving backward

There is no downtime in the management of high-powered science. New ideas or techniques need to be quickly exploited before scientists elsewhere do experiments that your people could have done first. Success automatically creates the need for new personnel and facilities that often require new buildings. If you are not ever agile in moving forward, the consequences will go beyond losing credit for the next breakthrough; top staff will move elsewhere to get the resources and support required to maintain their own leadership roles in their respective disciplines. As in sports, last year's championship doesn't count at the end of this season. Therefore, never be or expect other scientists to be sentimental about institutions. Going down with the ship is nuts.

4. Always buy adjacent property that comes up for sale

An institution's growth inevitably demands expanding the size of its campus. Therefore, never hesitate to buy land abutting yours even if there is no immediate use for it. Though the seller will charge you a premium knowing that land is worth more to someone with an adjacent lot than it is to a third party, your bargaining power is still greater than it will be when space is a dire need. It pays to overpay a little now rather than wait till your neighbor has you over a barrel.

5. Attractive buildings project institutional strength

Sometimes it seems money could be saved by building facilities that are less extravagant, with cheaper materials and unknown architects. This, however, is bad for business in the long term. Donors to Harvard never need fear that a building with their name on it will one day be sold off to stanch a hemorrhaging negative cash flow. Less than sturdy structures give out messages that their institution's life may be equally short. In contrast, solid, stylish buildings give donors the confidence that their descendants will one day bask in reflected glory. The lure of permanence inspires generosity.

6. Have wealthy neighbors

Research grants, no matter how much overhead they cover, are never enough to meet immediate needs. Unlike universities that can depend upon rich alumni, research institutions must have rich neighbors nearby who are inclined to take pride in local accomplishments. Typically their enthusiasm will be proportional to the research effort's potential to alleviate human misery. Nothing attracts money like the quest for a cure for a terrible disease.

7. Be a friend to your trustees

Entering worlds where your trustees relax—joining their clubs or vacationing where they go with their families in the summer, for instance—is a good way to put relations on a social footing. Seeing you as more friend than suppliant will incline them to go the extra distance for you in a pinch.

8. All take and no give will disenchant your benefactors

Philanthropy is a two-way street. The kindness of individuals donating to your cause should be acknowledged with a reciprocal gift to theirs. This by no means implies matching their generosity—it's the thought that counts. Since coming to Cold Spring Harbor, I have always made at least modest gifts to my trustees' other charities. By my so doing, they know I value the other activities that also give purpose to their lives. Equally important is being generous to your own institution. Why should others help if you don't also show you consider the cause worthy of some of your own discretionary income? It never hurts when those who decide your salary see you bite the bullet to give some of it back.

9. Never appear upset when other people deny you their money

Philanthropists are like others in not wishing to be taken for granted. Your institution is but one of many with its hand out. Soon after I became director, I approached the Bell Labs for sponsorship. My intermediary was a key scientist there who personally knew the value of our summer courses. When he called to say no money was coming our way, I expressed rude anger. Within minutes, I knew what a fool I had been to scotch any chance of our being shown some love in the future.

10. Avoid being photographed

You may raise the money but your success in the end totally depends upon those whose work you host, so it is their visibility, not yours, that matters. If they are not presentable, you are in deep trouble. Photos that put their names to their faces help them move more effectively in the outside world. They may not consider themselves photogenic, but they will enjoy the pride their spouses and especially their mothers will take in seeing them. Most important, keep your face out of your in-house publication unless you can be pictured beside a recognizable celebrity, such as Muhammad Ali or Lance Armstrong. Some of their glamour will momentarily rub off on you.

11. Never dye your hair or use collagen

A dye job works only if your hair is not noticeably thinning. It's impossible to seem as genuine as you must with a wan pate showing through a scattering of coal-black hairs. Of course, the impression of youthful vitality men sometimes cultivate with reddish dyes is even more disastrous. You wind up looking like Strom Thurmond. Gray hair and wrinkles at fifty bespeak dependability. Contrary to our present national ethos, it's better to act younger than you look, rather than the reverse. Harvard's Pusey had a wrinkleless face that reinforced the impression of a life devoid of pain or pleasure.

12. Make necessary decisions before you have to

Success more often comes from being first to take action than from being cleverer than your competitors. When it's the right thing to do, do it fast. If you wait too long, someone else is bound to propose it first and you will find yourself following someone else's agenda. In such circumstances, those taking orders from you will have reasonable cause to wonder why you are getting the top salary.

THIRTY years on, it would please me to report that the state of science at Harvard had righted itself in a manner befitting the world's richest and most influential university. The relatively short reign of Larry Summers as its twenty-seventh president, however, suggests that Harvard is once again headed in the wrong direction. Nothing may have distinguished Summers's time in office like the leaving of it, but his proposals for the future of science—which have yet to be modified by his successor—figured more critically in his vexed relations with the faculty than any clumsy words that marked his ultimate undoing. That the catalyst should have been his uttering an unpopular, though by no means unfounded, hypothesis only compounds the sorrows that any thinking person must feel contemplating the future of free inquiry at Harvard.

Despite Larry Summers's professed desire to move science onto Harvard's front burner, as his vision became clear many leading scientists on the Faculty of Arts and Sciences (FAS) were increasingly worried. Prominent among the uneasy was Tom Maniatis, who came back from Caltech in 1981 to become Professor of Molecular Biology. In his time at Harvard, Maniatis had been at the forefront of work in gene isolation and cloning, as well as a significant player in the biotechnology industry. And so it was notable when in the spring of 2003, he went to Massachusetts Hall to voice his concerns about Summers's plan to make Harvard the engine of a "second Silicon Valley," whose center would be a vast new campus of biology and medicine on

Harvard-owned land across the Charles River in Allston. The Allston vision was to be dominated by "translational research"—a term denoting science highly directed toward immediate application and, one might add, marketing. Tom was no stranger to the development of medical advances from cutting-edge science. He had, after all, founded with Mark Ptashne the successful biotech firm Genetics Institute and his lab had recently undertaken work on ALS. If anyone could appreciate a scientific utopia, it was he. In Tom's mind, however, Summers's plans would further weaken the already feeble heart of Harvard's historically distinguished pure biology programs still located in laboratories along Divinity Avenue to the immediate north of Harvard Yard.

Maniatis had barely begun to voice his opinions when Summers, oblivious to his visitor's distress, seized control of the conversation, using the remainder of the appointed hour to expound on his grand vision of how Harvard should move forward. He asked a purely rhetorical question: Should the president of Harvard be guided by the views of the perennial winners in the research game at Harvard or by the losers? By "the winners," Summers meant the Medical School, whose clinical studies had in recent decades brought the university its lion's share of awards and patents; "the losers" were the practitioners of basic science at FAS, which despite employing illustrious figures like Tom had lagged behind its rivals, most notably and proximately MIT. Summers would afterward inform a close aide that his meeting with Maniatis had gone extremely well; Tom meanwhile had walked out of Massachusetts Hall seething and more apprehensive than ever about the future of science at Harvard.

Only two years later Summers's inability to get outside his own head landed him in fatally hot water. It reached the boiling point following his appearance at a conference sponsored by the National Bureau of Economic Research on women in science, which was held in Cambridge in mid-January 2005. There he suggested that the relatively small number of women in tenured positions in the physical sciences might in part be attributable to a lower frequency among women as compared to men of innate potential for doing science at the highest level. Aware that many women would not take kindly to these words, he was careful to leave open the alternative explanation that in the past many talented women had been strongly discouraged by their teachers from ever trying to master top-level mathematics and sciences.

Summers's remarks might have gone unnoticed outside the meeting were it not for the presence of my former student, now a professor of biology at MIT, Nancy Hopkins. Over the past decade she had worked tirelessly and effectively to improve the working conditions of women scientists there. Before Nancy's highly visible efforts, the salaries and space assignments of women at MIT were notably unequal to those of their male counterparts. But Nancy did not challenge Summers at the meeting. Instead she instantly bolted from the room, later saying Summers's words made her sick, and soon appeared on national TV attacking him and setting off a firestorm of feminist anger.

It did Nancy Hopkins no particular credit as a scientist to admit that the mere hypothesis that there might be genetic differences between male and female brains—and therefore differences in the distribution of one form of cognitive potential—made her sick. Anyone sincerely interested in understanding the imbalance in the representation of men and women in science must reasonably be prepared at least to consider the extent to which nature may figure, even with clear evidence that nurture is strongly implicated. To my regret, Summers, instead of standing firm, within a week apologized publicly three times for being candid about what might well be a fact of evolution that academia will have to live with. Except for the psychologist Steve Pinker, no prominent Harvard scientist voiced a word in Summers's defense; I suspect the majority were fearful of being tarred with the brush of political incorrectness. If I were still a member of the faculty, the number of tenured scientists standing visibly behind the president in this matter would have literally doubled. But that would not have been enough to put out the flames. Apparently desperate, Summers soon contritely proposed a $50 million kitty to recruit more women to Harvard's senior science faculty.

The women-and-science firestorm by itself did not lead to Summers's dismissal late last February as Harvard's president. It was merely the culmination of hundreds of more private displays on his part of disregard for the social niceties that ordinarily permit human beings to work together for a common good. While academia almost expects its younger members to be brash and full of themselves, these qualities are most unbecoming in more seasoned members of the society, and generally fatal in leaders. Reading up on a topic the night before and then appearing at conferences with the bravado to suggest that one knows more than those who have spent their careers thinking about the issues at hand is no way for a president to act. Summers's non-age-adjusted IQ, moreover, at age fifty-one is likely 5 to 10 points lower than when he was a twenty-year-old wunderkind. Harvard's longstanding mandatory retirement age of fifty-five for academics was never a matter of arbitrary ageism but a recognition born of experience that as academics age they live more by old ideas and less by new ones. Summers, still acting as if he were the brightest person in the room, was bound to offend people who knew better.

It may be, however, that Summers is not entirely to blame for his social ineptitude. His repeated failures to comprehend the emotional states of those he presided over might be indicative of the genetic hand he was dealt as a mathematical economist—the very cards that endowed him with great quantitative intelligence may also have disabled the normal faculties for reading human faces and voices. The social incapacity of mathematicians is no mere stereotype; many of the most brilliant are mild to full-blown cases of Asperger's syndrome (the high-intelligence form of autism), perhaps the most genetically determined of known human behavioral "disabilities." Like exceptional math aptitude, Asperger's occurs five times more frequently in males than in females. The reason why must remain a mystery until further research shows how genes control the relative development and functioning of male and female brains.

If Summers's tactlessness does, in fact, have a genetic basis, much of the anger toward him should rightly yield to sympathy. No longer can his upbringing be blamed for failing to instill in him the graces of the civilized individual. In any case, all discussion should stop as to whether his dismissal was unduly precipitous—it was in all likelihood overdue. Whether those prominent individuals who promoted his candidacy should hang their heads in shame, however, is less obvious.

Summers's departure has to be seen as the first of many necessary steps to reclaim for Harvard its once legitimate claim to primacy in science, at least relative to MIT. Toward that end, Tom Maniatis, having come to Cold Spring Harbor to receive its honorary degree in April 2006, prevailed upon me to agree to meet with Derek Bok, the former president, whom the Harvard Corporation had called to serve again until a new leader could be found. A time was soon set up for me and Derek to get together at what had been the home of many Harvard presidents before Bok, the grand Georgian structure on Quincy Street across from the Faculty Club, now called Loeb House.

The day before my 10:00 A.M. appointment, Liz and I drove up to Cambridge to see Tom and his multitalented Long Island–born partner, Rachel von Roeschlaub. After an hour of tennis at the famous Longwood Cricket Club in Brookline, we drove in tandem to dinner at the Charles Hotel, where Liz and I were to stay the night. Before the food came, Tom described the brutal cuts in funding recently dealt out to Harvard's science departments when Summers had launched a massive hiring campaign in translational life sciences and committed large pots of cash to develop Allston.

Derek's initial reaction to my furrowed face must have been the same as mine to his graying hairs. Neither of us could pass for middle-aged. Twenty-eight years had elapsed since we were last face to face at the June 1978 commencement ceremonies, on which occasion I was delighted to be receiving an honorary D.Sc. A photo from that day shows Derek at forty-eight and me at fifty, each happy with himself, with a few more good years left to advance the causes of Harvard (in his case) and of scientific research (in mine). Neither of us could then have suspected that nearly thirty years later we might yet have cause to meet in the service of both.

Knowing that a petitioner's allotted hour always passes quickly, I went straight to my main message: It was wrongheaded to build a huge Allston biology complex to compensate for the lack of greatness increasingly enveloping the biology labs along Divinity Avenue. Most likely, I argued, it was the B+ level of most of Harvard's life sciences, both in Cambridge and across the vast medical complex, that would gravitate to the brochure-perfect new campus, resulting in very little

*With Derek Bok
at Harvard
commencement
ceremony, June 1978*

bang for the vast bucks that would be spent. This, I insisted, is not how great institutions are built.

I went on to tell Derek that he and the Harvard Corporation should ask why MIT's life sciences now so completely outclass Harvard's. Past stinginess of Harvard deans played a big role in a problem that indiscriminate lavishness could not now fix. For far too long, University Hall had witlessly acted as if Harvard did not have to spend its own money to keep a place in the top league of science. The leadership assumed that Harvard's golden name would naturally move the federal government to fund not only the university's research but also the creation of new facilities required to stay at the cutting edge. But brand names count for very little in science. And so, foolishly, Harvard sat back on its heels for about two decades while MIT smoothly integrated the privately funded Whitehead Institute into its biology operations and, under the never shy Eric Lander, created a huge DNA sequencing facility. Thus MIT became a major player in the Human Genome Project, the intellectual driveshaft for much of today's most exciting biology and medicine.

Only belatedly had Harvard tried to enter the Genome Age by committing itself, as the twenty-first century began, to becoming strong in systems biology, a discipline so sprawling and unwieldy as to merit comparison to Enron's limitless expansions before its collapse into nothingness. In turn, the large MIT Center for Genome Research, thanks to the generosity of the California-based philanthropists Eli and Edythe Broad, was able to metamorphose in 2003 into an even more ambitious incarnation, the Broad Institute of MIT and Harvard. Its sleek Cambridge Center edifice across from MIT's Koch Biology Building would have looked as much at home in the Boston financial district. By diverting funds that might have been spent along Divinity Avenue, Harvard under Summers bought a say in how and by whom the Broad's massive genomic resources would be utilized. Within its doors, however, Harvard Yard seems light-years away.

Through such lavish commitments to joint ventures the pain of being shortchanged continued to be felt along Divinity Avenue. Still much rued was the failure in 2001 of the Faculty of Arts and Sciences to lure the clever Roderick MacKinnon, then in his late forties, away from Rockefeller. Longtime Harvard stalwart and X-ray structure whiz Steve Harrison, who was convinced that MacKinnon's crystallographic studies on ion channels would earn him a Nobel Prize (he would indeed share the 2003 award in Chemistry), felt likewise certain that with the right inducement MacKinnon would return to Harvard, where he'd had a lab at the medical school before leaving for Rockefeller. The package offered to him by then-dean Jeremy Knowles, however, was not remotely competitive with Rockefeller's commitment. Upon seeing the dean's letter, in fact, MacKinnon's wife wondered whether a mistake had been made in locating the decimal point. Depressed at Knowles's failure to think realistically big, Steve himself developed a case of wanderlust and spied a much brighter future for himself at Harvard Medical School. He wasted little time moving his highly productive X-ray crystallographic research group across the Charles.

Several years before these events, a dinner party at Mark and Lucy Ptashne's spectacularly renovated huge house on Sparks Street had reunited me and Jeremy, whom I first knew when he was one of Oxford's stars in chemistry. Given his background, I had assumed he would use his new powers as Harvard's number two to brighten the future of science there. So I was slack-jawed as Jeremy told the assembled scientists and their spouses of the forthcoming boon to their work in the form of $1 million for supplies and equipment he would soon disburse among all the science departments. I blurted out that such a pittance would scarcely cover a small fraction of the scientists working at Cold Spring Harbor, adding that the miserly way Oxford was being run toward insignificance was no way for Harvard to keep pace with MIT. The stunned silence made me realize that no one had ever before witnessed such brazen disrespect for University Hall. Back at the Charles Hotel, I went to bed imagining Jeremy moving through a Max Beerbohm short story.

At this writing, University Hall is still under temporary stewardship. Larry Summers's firing of the East Asian studies scholar Bill Kirby as dean was not only the last straw of the Summers presidency; it also paralyzed Harvard administratively. But making new faculty appointments is not something that can wait—a short hiatus from hiring could do years of damage. Derek correctly understood that the job must be filled at once but also that the next president had the right to choose the next dean. So just prior to our appointment, Derek asked Jeremy to return temporarily to the helm of the Faculty of Arts and Sciences.

We now know that the eminent historian Drew Gilpin Faust will be the next president of Harvard. Who should the next dean be? Clearly it must be someone of commanding intellect and deep knowledge of the

Harvard scene. But even if he or she also possesses Henry Rosovsky's uncanny sense of knowing when not to say no, this person will be taking on a role now too large for one individual. For the sake of excellence in all areas of inquiry, Harvard should divide the responsibility into three more manageable groupings—science, humanities, and social sciences. Each should be led by a distinguished academic with substantial powers of the purse. Only the capacity to judge and pay market rates can assure a better than 50-50 chance that the first-choice candidates of the respective ad hoc committees will accept a Harvard offer.

To be certain of success overall, Harvard salaries must once again be much higher than those of serious competitors. To get stars, you must offer star salaries. The best of academia no longer will come to Harvard because it is Harvard. No one goes into scientific research to get rich, but one doesn't undertake it to evade the comforts of life. Living close enough to Harvard Yard to enjoy its ambiance and diversions is now beyond the means of new Harvard appointees with families unless the faculty salary is matched by another of the same magnitude. Paying top salaries is well within the means of the largest university endowment on earth—provided that it abandon the almost Soviet-style fantasy of the Allston expansion, at present envisioned to cover the area of twenty-five football stadiums.

Science that leads over the horizon depends on gathering the best minds and enabling them to do what the best minds naturally seek to do: pursue the most thrilling questions of the time. Such minds inevitably draw their like, and the rest takes care of itself. The dividends of such greatness, however, go beyond what is to be gained by winning the next scientific race. They extend to the enrichment of the student body by giving them a broader appreciation of intellectual values.

Harvard's new president will need to see paramount among her goals the seeking of potential greatness for its undergraduates through equipping them with the best ideas of the past, honest assessments of the world today, and realistic expectations about the future. This was Robert Hutchins's vision for the University of Chicago when in 1929, at age thirty-one, he became its president. His charismatic impact reached its apogee in the 1940s as great books and ideas became the mainstays of undergraduate education along the Midway. Though his successors, pressed to maintain a viable inflow of new undergraduates, saw the need for partial retreat from the purity of his vision, those educated in the primacy of great thoughts never doubted that they were the chosen people.

Even now, mention of the University of Chicago to educational leaders not directly exposed to Hutchins or to his immediate successors elicits wistful admissions that Hutchins largely had it right when he branded much of American higher education as a prolonged mismatch of triviality and ignorance. During our meeting Derek allowed that despite the failure of Hutchins's ideas to take hold at any other major American university, his was the only past American university presidency that educators still actively talk about. While the Ivy League turns out graduates who for the remainder of their lives seek out one another, the University of Chicago still strives to see its graduates leave with lifetime-long ideas and a passion to see the world as it is. Parrington's *Main Currents in American Thought* and *The Brothers Karamazov* had much more impact on my life than any of my University of Chicago classmates.

At their best, universities promote outside their walls the spirit and values that enable the proper conduct of their work within those walls. Going to the College of the University of Chicago completed my conversion to a life devoted to discovery of the natural world for its own sake, the impulse first stirred by looking for birds with my father. Yet whatever great advance in knowledge a university may bring us, it will fail in its ultimate mission if it allows concerns such as self-marketing and customer satisfaction—concerns of the service institutions that most universities are fast becoming—to overtake the pure good of pursuing truth. And this is particularly important to science, in which the race, though it may be to the swift, is never over.

Before leaving Derek's temporary office I remarked that the time was surely not far off when academia would have no choice but to hand political correctness back to the politicians. Since 1978, when a pail of water had been dumped over E. O. Wilson for saying that genes influence the behavior of humans as well as of other animals, the assault against behavioral science by wishful thinking has remained vigorous. But as science is able to better prove its hypotheses, such irrationality must recede or betray itself as such. In showing that human genes do matter, behavioral biologists will no longer be limited to comparisons of fraternal and identical twins. Soon the cost of sequencing the A's, T's, G's, and C's of individual DNA molecules will drop to a thousandth of what it has been, thereby transposing our studies of behavioral differences to the much more revealing molecular level. DNA messages extracted from, say, many hundreds of psychopaths can then be compared to equivalent numbers of DNA messages from individuals prevented by their consciences from habitually lying, stealing, or killing. Specific DNA sequences consistently occurring only in psychopaths will allow us to pinpoint the genes whose malfunctions are likely to produce psychopathy. The thought that some people might be born to grow up wicked is inherently upsetting. But if we find such behavior to be innate, the integrity of science, no less than that of ethics, demands that we let the truth be known.

The relative extents to which genetic factors determine human intellectual abilities will also soon become much better known. At the etiological heart of much of schizophrenia and autism are learning defects resulting from the failure of key brain cells to link up properly to each other. As we find the human genes whose malfunctioning gives rise to such devastating developmental failures, we may well discover that sequence differences within many of them also lead to much of the observable variation in human IQs. A priori, there is no firm reason to anticipate that the intellectual capacities of peoples geographically separated in their evolution should prove to have evolved identically. Our wanting to reserve equal powers of reason as some universal heritage of humanity will not be enough to make it so. Rather than face up to facts that will likely change the way we look at ourselves, many persons of goodwill may see only harm in our looking too closely at individual genetic essences. So I was not surprised when Derek asked apprehensively how many years would pass before the key genes affecting differences in human intelligence would be found. My back-of-the-envelope answer of "fifteen years" meant Summers's then undetermined successor would not necessarily need to handle this very hot potato.

Upon returning to the Yard, however, I was not sure that even ten years would pass.

Cast of Characters

*With Swedish pro
Carl Wermee
at Piping Rock Club in
Locust Valley, New York*

George Beadle (1903–1989)—After heading the Biology Division at Caltech from 1946, in 1961 he became the president of the University of Chicago, so serving until he was sixty-five. Then, as director of the Institute for Biomedical Research of the American Medical Association, he resumed research on the origins of modern corn. In 1982 he moved with his wife, Muriel, a writer, to a retirement village in Pomona, California.

Seymour Benzer (b. 1921)—In 1976, he moved from Purdue to Caltech, where he exchanged phage for *Drosophila,* using it to effectively probe the genetic basis of behavior and neurodegeneration.

Derek Bok (b. 1930)—After retiring as Harvard president in 1991, he remained highly involved with higher education, writing six books on the topic: *Our Underachieving Colleges* (2005), *Universities in the Marketplace* (2003), *The Shape of the River* (1998), *Universities and the Future of America* (1990), *Higher Learning* (1986), and *Beyond the Ivory Tower* (1982). His recent research focuses on the U.S. government's approach to domestic problems, about which he has written two books, *The State of the Nation* (1997) and *The Trouble with Government* (2001). Following the resignation of Larry

Summers, Derek returned in July 2006 to Harvard to serve as acting president for one year.

Sir (William) Lawrence Bragg (1890–1971)—He left his post as director of the Cavendish Laboratory at Cambridge University in 1954 to head the Royal Institution in London, which his father, William Henry Bragg, had directed between 1930 and 1942.

Sydney Brenner (b. 1927)—Upon the retirement of Max Perutz, he became head of the Laboratory of Molecular Biology at Cambridge, serving until 1986. By then he was devising methodologies for studying the human genome, early on seeing the importance of making DNA copies of cellular messenger RNA molecules. Increasingly he worked outside the United Kingdom, at the Scripps Research Institute in La Jolla, California, at the Molecular Sciences Institute that he founded in Berkeley, and in Singapore as a biotechnology adviser to its government. In the United Kingdom, he and his wife, May, maintain their primary residence in the town of Ely, to the north of Cambridge.

Jacob (Bruno) Bronowski (1908–1974)—One of the early research fellows of the Salk Institute for Biological Studies, he later spent two years (1971–1972) filming the justly famous BBC series *The Ascent of Man,* which traced the history of science and mankind from prehistoric times and which aired just shortly before his tragically premature death from heart failure.

McGeorge Bundy (1919–1996)—In 1966 he left the Lyndon Johnson White House to direct the Ford Corporation in New York for twelve years. During the following ten years he taught history at New York University, subsequently becoming a scholar-in-residence at the Carnegie Corporation, where he chaired its Committee on Reducing Nuclear Dangers.

Dick Burgess (b. 1942)—Following two years as a postdoctoral fellow in Geneva, Switzerland, he joined the faculty of the University of Wisconsin, Madison, where he is currently the James D. Watson Professor of Oncology.

John Cairns (b. 1922)—After leaving Cold Spring Harbor Laboratory in 1972, he headed the Mill Hill Laboratory of the Imperial Cancer Research Fund outside London until 1980. He then recrossed the Atlantic to join the Harvard School of Public Health. Upon his retirement in 1991, he and his wife, Elfie, moved back to the United Kingdom, living outside Oxford.

Mario Capecchi (b. 1937)—After finishing his Ph.D. work, he stayed on at Harvard as assistant and then associate professor of biochemistry until 1973, when he moved to the University of Utah, where he has remained since. There he has pioneered gene targeting in mouse embryo–derived stem cells.

Erwin Chargaff (1905–2002)—More a writer than a scientist in the later part of his career, he published several books, including the autobiographical *Heraclitean Fire: Sketches from a Life Before Nature.* He remained on the Columbia faculty until his retirement in 1974.

Seymour Cohen (b. 1917)—In 1971, he left the University of Pennsylvania for the University of Colorado in Denver, where he was a professor in the School of Medicine until 1976. He then moved to the State University of New York at Stony Brook, from which he retired in 1985.

Francis Crick (1916–2004)—At the age of sixty-one, he moved to the Salk Institute to pursue a new career as a neurobiologist, eventually to study the nature of consciousness with Caltech's Christof Koch. The already much valued biography, *Francis Crick: Discoverer of the Genetic Code* (Atlas Books/ Harper Collins), by English scientist/writer Matt Ridley, appeared in mid-2006.

Manny Delbrück (1917–1998)—She continued to live at Caltech until her death from breast cancer.

Max Delbrück (1906–1981)—After 1957, his research turned to problems in sensory physiology, which he studied using the mold *Phycomyces,* until his death from multiple myeloma. In 1969, he received the Nobel Prize in Physiology or Medicine together with Alfred Hershey and Salvador Luria for their work on bacteriophage.

Milislav Demerec (1895–1966)—Following his retirement at age sixty-five as director of Cold Spring Harbor Laboratory in 1960, he continued work on *Salmonella* genetics at Brookhaven National Laboratory until 1965.

August (Gus) Doermann (1918–1991)—After conducting research at Oak Ridge National Laboratory, he worked at Rochester and Vanderbilt before becoming professor of genetics at the University of Washington in 1964. His retirement years after 1982 were spent in the Canadian Yukon.

Paul Doty (b. 1920)—In his later career, he became increasingly involved in issues of international security, founding in 1973 what is now the Belfer Center for Science and International Affairs at the John F. Kennedy School of Government at Harvard.

Renato Dulbecco (b. 1914)—After leaving Indiana University for Caltech in 1949, he began research on animal viruses, eventually to focus on tumor viruses as a founding member of the Salk Institute. His lab's finding that DNA tumor viruses cause cancer by inserting their genes into host cell DNA led to his sharing the 1975 Nobel Prize in Physiology or Medicine. Later he became one of the earliest proponents of the Human Genome Project, saying that until the human genome was known we would not have the knowledge to beat cancer.

Julian Fleischman (b. 1933)—After finishing his Ph.D. at Harvard in 1960, he went on to postdoctoral training at Stanford University and several other institutions. In 1963 he coauthored a paper proposing a detailed structure for the antibody molecule. He later joined the Department of Molecular Microbiology at Washington University School of Medicine in St. Louis, where he continues to study antibody structure, synthesis, and diversity.

Rosalind Franklin (1920–1958)—After moving to Birkbeck College in the spring of 1953, she worked on the structure of tobacco mosaic virus until her tragically premature death from ovarian cancer in 1958. Her life is the subject of Brenda Maddox's much acclaimed biography *Rosalind Franklin: The Dark Lady of DNA*.

Carleton Gajdusek (b. 1923)—He spent the majority of his career at the National Institutes of Health and received the 1976 Nobel Prize in Physiology or Medicine for his research on the infectious nature and epidemiology of the prion disease kuru, which he studied in a population of South Fore people in the highlands of New Guinea beginning in the mid-1950s.

George (Geo) Gamow (1904–1968)—After a twenty-two-year career as professor of physics at George Washington University, he moved to the University of Colorado at Boulder in 1956, where he worked until his death.

Ray Gesteland (b. 1938)—After spending his postdoctoral years in Geneva with Alfred Tissières, he came in 1967 to Cold Spring Harbor Laboratory, where he served as assistant director. In 1978 he moved to the University of Utah, where he is now vice president for research.

Celia Gilbert (b. 1932)—She now divides her time between painting, writing poetry, and taking delight in her children and grandchildren.

Wally Gilbert (b. 1932)—He received the 1980 Nobel Prize in Chemistry with Frederick Sanger for their independent development of DNA sequencing methods. In 1982, he briefly left Harvard to run Biogen, the then Swiss-based biotechnology company he had helped found two years earlier. No longer running a research group as professor emeritus, he remains a senior fellow of the Harvard Society of Fellows and continues to be closely involved with biotechnology. He also devotes much time to photography and classical antiquities.

Don Griffin (1915–2003)—After leaving Harvard's Biology Department in 1965, he moved on to Rockefeller University's field station for behavioral studies in Millbrook, north of New York City.

Gary Gussin (b. 1939)—After his postdoctoral years in Geneva, Switzerland, he joined the faculty of the University of Iowa, where he is currently professor of biological sciences.

Alfred Hershey (1908–1997)—In 1969 he shared the Nobel Prize in Physiology or Medicine for his and Martha Chase's 1950 demonstration that phage DNA, not protein, is its genetic material. He retired in 1972 from active research at Cold Spring Harbor Laboratory, continuing to live near the lab until his death.

Nancy Hopkins (b. 1943)—After joining the Center for Cancer Research at MIT in 1973, she pursued research on RNA tumor viruses. Switching her focus later to zebra fish, she developed a new method of insertional mutagenesis that identified hundreds of genes necessary for zebra fish development. In recent years she has promoted gender equity at MIT, where she is now the Amgen Professor of Biology.

Robert Hutchins (1899–1977)—After leaving the University of Chicago in 1951, he became associate director of the Ford Foundation and the chairman of its new Fund for the Republic. In 1959 he founded the Center for the Study of Democratic Institutions in California, which he led until his death.

François Jacob (b. 1920)—After being awarded the 1965 Nobel Prize in Physiology or Medicine with André Lwoff and Jacques Monod for their work on bacterial gene regulation, he has continued to work at the Institut Pasteur, where he served as chairman of the board from 1982 to 1988. Among his influential books are his autobiography, *The Statue Within; The Logic of Life;* and more recently *Of Flies, Mice, and Men.*

Herman Kalckar (1908–1991)—In 1952 he returned to the United States to work first at the National Institutes of Health, then at Johns Hopkins University, and finally, in 1961, at Harvard Medical School, as head of the Biochemical Research Laboratory of Massachusetts General Hospital.

John Kendrew (1917–1997)—In the early 1970s he helped create the European Molecular Biology Laboratory in Heidelberg, becoming its first director when it opened in 1974. He then served as president of St. John's College, Oxford, from 1981 to 1987.

Charles Kurland (b. 1936)—After receiving his Ph.D. from Harvard, he moved on to a postdoctoral position at the Microbiology Institute of the University of Copenhagen. From there he moved to the University of Wisconsin and then in 1971 to Uppsala University in Sweden. He is now professor emeritus of its Department of Molecular Evolution as well as at the Department of Microbiology at the University of Lund.

Joshua Lederberg (b. 1925)—In 1959 he left the University of Wisconsin to found and chair the Department of Genetics at Stanford University; in 1978 he moved again to Rockefeller University in New York, where he served as president until his retirement in 1990. In 2006, he received the Presidential Medal of Freedom

Salvador Luria (1912–1991)—He moved in 1959 from the University of Illinois to the Biology Department of MIT. In 1972, he was asked to plan the new Center for Cancer Research, an endeavor that involved remodeling a former chocolate factory into a laboratory building. He was its director from its opening in 1973 until 1985. In 1969 he shared the Nobel Prize in Physiology or Medicine with Max Delbrück and Alfred Hershey.

Ole Maaløe (1915–1988)—He moved from the State Serum Institute to the University of Copenhagen's Department of Microbiology, where he remained until his retirement.

Tom Maniatis (b. 1943)—He is currently the Thomas H. Lee Professor of Molecular and Cellular Biology at Harvard University, where his lab has in recent years focused on the regulation of eukaryotic gene expression, particularly in the immune response. In 1982 he co-founded, with Mark Ptashne, the biotech company Genetics Institute, now part of Wyeth Pharmaceuticals, and more recently Acceleron Pharma.

Ernst Mayr (1904–2005)—In 1975 he retired from the Harvard University faculty under the title Alexander Agassiz Professor of Zoology Emeritus. He went on to publish some two hundred articles and fourteen books between his official retirement and his death at the age of 100.

Barbara McClintock (1902–1992)—She became the sole recipient of the 1983 Nobel Prize in Physiology or Medicine, awarded for her work at Cold Spring Harbor Laboratory on the transposition of maize genes. Though she formally retired from the Carnegie Institution's Department of Genetics in

1967, she remained an important presence at the laboratory until her death at Huntington Hospital, near Cold Spring Harbor, at the age of ninety.

Matt Meselson (b. 1930)—He is currently the Thomas Dudley Cabot Professor of the Natural Sciences at Harvard University. Since the 1960s he has been highly involved in chemical and biological weapons and arms control, and is now director of the Harvard Sussex Program and Chair for CBW studies at the Belfer Center for Science and International Affairs at the Kennedy School of Government at Harvard.

Avrion (Av) Mitchison (b. 1928)—In 1970, he left Mill Hill to become professor of zoology at University College, London, continuing to focus on immunology. Following his retirement from UCL, he worked for several years in Berlin before returning to London.

Naomi (Nou) Mitchison (1897–1999)—She continued to write well into her eighties, and died at Carradale, in western Scotland, at the age of 101. By then her life had been seriously compromised by Alzheimer's disease.

Jacques Monod (1910–1976)—He shared the 1965 Nobel Prize in Physiology or Medicine with André Lwoff and François Jacob. His book-length essay *Chance and Necessity,* explaining life and evolution as a result of chance, was first published in 1970. Though he held positions at the University of Paris, Collège de France, and the Salk Institute, he centered his career at the Institut Pasteur, serving as its director from 1971 until his death of leukemia at his home in Cannes, France, at the age of sixty-six.

H. J. Muller (1890–1967)—He continued working with *Drosophila* at Indiana University for the remainder of his career.

Benno Müller-Hill (b. 1933)—In 1968, two years after isolating the lactose repressor with Wally Gilbert, he became a professor at the Institute for Genetics of the University of Cologne. In 1984, he published his revealing description of German eugenics, *Todliche Wissenschaft,* which sold fifteen thousand copies in Germany but received only one review in the German press. In 1988 its English translation, *Murderous Science: Elimination by Scientific Selection of Jews, Gypsies, and Others in Germany, 1933–1945,* was published by Oxford University Press, with a later 1998 edition published by Cold Spring Harbor Laboratory Press.

Betty (Watson) Myers (1930–1999)—She continued to live in Washington upon her husband Robert Myers's leaving government service to become publisher of *The Washingtonian.* In 1980 they moved to New York City, where he became president of the Carnegie Council on Religion and International Affairs, and they resided there until 1996, when they moved to Menlo Park, California, to be close to their daughters.

Masayasu Nomura (b. 1927)—In 1984, after twenty years on the faculty at the University of Wisconsin, Madison, he moved to the University of California, Irvine, where he still runs an active research group that studies RNA synthesis in the yeast *Saccharomyces cerevisiae.*

Aaron Novick (1919–2000)—He moved from the University of Chicago to the University of Oregon in 1959 to become the founding director of its Institute of Molecular Biology. At Oregon, he went on to serve as dean of the Graduate School and head of the Biology Department.

Linus Pauling (1901–1994)—He left Caltech shortly after winning the 1962 Nobel Peace Prize for his activism against nuclear testing and proliferation. He spent the next decade at the Center for the Study of Democratic Institutions in Santa Barbara, California (1963–1967); the University of California, San Diego (1967–1969); and Stanford University (1969–1974). In 1973 he founded the Linus Pauling Institute of Science and Medicine in Palo Alto, California, where he conducted research on the effects of common micronutrients on human health. He died of cancer at his ranch in Big Sur, California.

Peter Pauling (b. 1931–2003)—He was a lecturer in physical chemistry at University College London from 1958 to 1989. After his retirement he lived in Wales until his death.

Max Perutz (1914–2002)—He served as chairman of the MRC Laboratory of Molecular Biology at Cambridge University from its inception in 1962 until 1979, the year he officially retired. He continued to work at the Laboratory practically every day after his retirement until his death of cancer at the age of eighty-seven. Increasingly he was celebrated as a skilled writer, publishing many articles in *The New York Review of Books.* Collections of his essays were also published in book form: *Is Science Necessary?* (Oxford University Press, 1991) and *I Wish I'd Made You Angry Earlier* (Cold Spring Harbor Laboratory Press, 1998 and 2003).

Ulf Pettersson (b. 1942)—He is currently vice rector of the Disciplinary Domain for Medicine and Pharmacy and professor in the Department of Genetics and Pathology at Uppsala University in Sweden.

Princess Christina (b. 1943)—She married Tord Gösta Magnuson in 1974 and became known as Princess Christina, Mrs. Magnuson. She currently lives in Stockholm.

Mark Ptashne (b. 1940)—After isolating the lambda phage repressor in 1967, his continued research yielded a detailed picture of genetic regulation in lambda. More recently he has focused on transcriptional regulation in yeast. In 1997, he moved to the Memorial Sloan-Kettering Cancer Center in New York City, where he is currently the Ludwig Chair of Molecular Biology.

Edward Pulling (1898–1991)—He continued to live at Redcote after the death of his wife, Lucy, in 1979, remaining as chairman of the Long Island Biological Association until early 1986. Afterward, he took much pleasure from writing about earlier years in his life.

Nathan Pusey (1907–2001)—After retiring early from Harvard in 1971, he moved to New York City to serve for several years as president of the Andrew W. Mellon Foundation.

J. T. (Sir John) Randall (1905–1984)—In 1970 he retired as the first director of the Biophysics unit at King's College, London, which was soon to be renamed the Randall Institute and is now the Randall Division of Cell and Molecular Biophysics.

Alex Rich (b. 1924)—He spent four years at the National Institutes of Health before moving in 1954 to MIT, where he continued work on nucleic acids, in particular their 3-D structures. He was honored with the National Medal of Science in 1995 and still maintains an active research career as MIT's William Thompson Sedgwick Professor of Biophysics.

John Richardson (b. 1938)—After post-doctoral research at the Institut Pasteur, he joined the faculty of Indiana University, Bloomington, where his research continues to focus on RNA synthesis from DNA templates.

Bob Risebrough (b. 1935)—After Harvard, he pursued a career as a conservation biologist. He lives in California, working at the Bodega Bay Institute and most recently focusing his research on the endangered California condor.

Keith Roberts (b. 1945)—He remains at the John Innes Centre in Norwich, where he served for ten years as head of its Cell Biology Department. In 1983 he helped author the widely used *Molecular Biology of the Cell*, now in its fifth edition.

Rich Roberts (b. 1943)—He shared the 1993 Nobel Prize in Physiology or Medicine with Phil Sharp for their independent 1977 discovery of RNA splicing, using adenovirus DNA. While at Cold Spring Harbor Laboratory, his lab discovered some one hundred new restriction enzymes. In 1992 he moved to New England Biolabs, a major provider of research reagents, where he is currently chief scientific officer.

Charles S. Robertson (1905–1981)—He was stricken with Alzheimer's disease in the late 1970s, and died in 1981 at his home in Delray Beach, Florida.

Henry Rosovsky (b. 1927)—He stepped down as dean of the Faculty of Arts and Sciences at Harvard in 1984, and in 1996 retired as the Lewis P. and Linda L. Geyser University Professor Emeritus. In 1990 he published *The University: An Owner's Manual,* a book about his experiences in the world of higher education.

Jonas Salk (1914–1995)—In his last decade, his focus was increasingly on AIDS and the need for an effective vaccine against it. He directed the Salk Institute until his death.

Joe Sambrook (b. 1939)—He left Cold Spring Harbor Laboratory in 1985 to head the department of biochemistry at the University of Texas Southwestern Medical Center. Ten years later, he moved to the Peter MacCallum Cancer Centre in Melbourne, Australia, where he has played an active role in advancing breast cancer research. In 2006 he was appointed Executive Scientific Director of the Australian Stem Cell Centre.

David Schlessinger (b. 1936)—After receiving his Ph.D. from Harvard, he completed his postdoctoral training at the Institut Pasteur and then joined Washington University in St. Louis, eventually to serve as professor of molecular microbiology. In 1997 he moved to the National Institute on Aging in Baltimore, Maryland, where he is currently Chief of its Laboratory of Genetics.

Phil Sharp (b. 1944)—After leaving Cold Spring Harbor, he spent the rest of his career at MIT where he has headed its Center for Cancer Research (1985–1991), the Department of Biology (1991–1999), and the McGovern Institute for Brain Research (2000–2004). In 1993 he shared the Nobel Prize in Physiology or Medicine with Rich Roberts for their independent discovery of RNA splicing. He cofounded Biogen in 1978 and, more recently, Alnylam Pharmaceuticals (2002).

Tracy Sonneborn (1905–1981)—He spent the rest of his career at Indiana University, Bloomington; he retired from teaching in 1976 but continued to do research until his death at the age of seventy-five.

Gunther Stent (b. 1924)—In 1952 he moved from Caltech to the University of California at Berkeley, where he has remained since. During a 1972 sabbatical at Harvard, be began studying neurobiology, and soon developed a leech colony at Berkeley to support his lab's neuroembryology and neurophysiology research. He is currently professor emeritus of neurobiology.

Leo Szilard (1898–1964)—He remained loosely affiliated with the University of Chicago until he became a founding fellow of the Salk Institute in 1963. Only several months after moving there, he died of a heart attack and his ashes, at his wish, were dispersed over the Pacific Ocean.

Howard Temin (1934–1994)—Howard remained at the University of Wisconsin for his entire academic career. In 1975, with Renato Dulbecco and David Baltimore, he was awarded the Nobel Prize in Physiology or Medicine for research on tumor viruses. Though never a smoker, he became a victim of lung cancer, dying at the early age of fifty-nine.

Alfred Tissières (1917–2003)—He remained at the Department of Molecular Biology of the University of Geneva for the rest of his career, living with his wife, Virginia, in the nearby village of Vandœuvres. A sabbatical year at Caltech led to his influential work on the heat shock response in *Drosophila*.

Alex Todd (1907–1997)—After his work on the structure and synthesis of nucleotides, which won him the 1957 Nobel Prize in Chemistry, he continued to work on the structures of a variety of molecules found in insects and plants. He retired from his positions as professor of organic chemistry at Cambridge University in 1971 and lived in Cambridge until his death.

Niccolò Visconti di Modrone (1920–2004)—After working between 1950 and 1954 on phage genetics with Al Hershey at Cold Spring Harbor, Niccolò left research for business and cofounded the pharmaceutical company Lepetit, and later Pierrel. Upon retiring, he took up residence in his hometown of Milan and maintained a close friendship with his colleagues at CSHL, including Waclaw Szybalski, Evelyn Witkin, and Barbara McClintock.

Liz (Lewis) Watson (b. 1948)—She has received master's degrees in historic preservation from Columbia University and in library and information science from Long Island University. She has devoted much time to preserving the historic buildings of Cold Spring Harbor Laboratory and in 1991 published a pictorial history of CSHL titled *Houses for Science.* For many years she was also a member of the Huntington Historic Preservation Commission, and in 2001 she was appointed by Governor George Pataki to the New York State Board for Historic Preservation.

Klaus Weber (b. 1936)—Since 1975 he has worked at the Max Planck Institute for Biophysical Chemistry in Göttingen, Germany, where he is now professor emeritus. Along with wife Mary Osborn, he pioneered the use of immunofluorescence microscopy and has made significant contributions to the study of the cytoskeleton.

Jerome Wiesner (1915–1994)—He served as president of MIT from 1971 to 1980 and devoted a majority of his time after retirement to teaching and policy research, strongly focusing on nuclear arms control.

Maurice Wilkins (1916–2004)—After his contributions to the double helical structure of DNA at King's College, London, he focused increasingly on issues of the social responsibility of scientists, teaching undergraduate courses on the topic and serving for twenty-two years as president of the British Society for Social Responsibility in Science. He was also very proud of his involvement in the Campaign for Nuclear Disarmament.

Carroll Williams (1916–1991)—He spent his entire career in insect physiology at Harvard, starting with the completion of both his Ph.D. and M.D. and his appointment to its faculty in 1946, from which he retired in 1987.

Edward O. Wilson (b. 1929)—He is currently Pellegrino University Professor Emeritus and curator of entomology at the Museum of Comparative Zoology at Harvard University. In recent years his interests have turned increasingly to conservation biology.

Tom Wilson (1902–1969)—He died of a heart attack all too soon after his move in 1968 from Harvard University Press to Atheneum Publishers in New York City.

Barbara Wright (b. 1926)—She continued in research after marrying Herman Kalckar and moving to the United States, working first at NIH and later at Massachusetts General Hospital and the Boston Biomedical Research Institute. In 1982 she became a research professor in biological sciences at the University of Montana.

Sewall Wright (1889–1988)—After his mandatory retirement from the University of Chicago at the age of sixty-five, he moved to the University of Wisconsin where he served as professor of genetics for thirty-four more years.

Norton Zinder (b. 1928)—Continuing study of phage f1 and its interactions with *E. coli*, he remained all his career at the Rockefeller University where he is now professor emeritus.

David Zipser (b. 1937)—After completing a postdoc with Sydney Brenner and then serving on the faculty of Columbia University, he moved in 1970 to Cold Spring Harbor Laboratory as its bacterial geneticist. Increasingly attracted by brain research, he left CSH in 1982 for the University of California, San Diego, where he is now Professor Emeritus of Cognitive Neuroscience.

Remembered Lessons

Chapter 1. FROM CHILDHOOD ON CHICAGO'S SOUTH SIDE

1. Avoid fighting bigger boys or dogs
2. Put lots of spin on balls
3. Never accept dares that put your life at risk
4. Accept only advice that comes from experience as opposed to revelation
5. Hypocrisy in search of social acceptance erodes your self-respect
6. Never be flippant with teachers
7. When intellectually panicking, get help quickly
8. Find a young hero to emulate

Chapter 2. FROM ADOLESCENCE AT THE UNIVERSITY OF CHICAGO

1. College is for learning how to think
2. Knowing "why" (an idea) is more important than learning "what" (a fact)
3. New ideas usually need new facts
4. Think like your teachers
5. Pursue courses where you get top grades
6. Seek out bright as opposed to popular friends
7. Have teachers who like you intellectually
8. Narrow down your intellectual (career) objectives while still in college

Chapter 3. FROM YOUNG ADULTHOOD AT INDIANA UNIVERSITY

1. Choose a young thesis adviser
2. Expect young hotshots to have arrogant reputations
3. Extend yourself intellectually through courses that initially frighten you
4. Humility pays off during oral exams
5. Avoid advanced courses that waste your time
6. Don't choose your initial thesis objective
7. Keep your intellectual curiosity much broader than your thesis objective

Chapter 4. From summering at Cold Spring Harbor

1. Use first names as soon as possible
2. Banal thoughts necessarily also dominate clever minds
3. Work on Sundays
4. Exercise exorcises intellectual blahs
5. Late summer experiments go against human nature

Chapter 5. From observing Leo Szilard and Max Delbrück

1. Have a big objective that makes you feel special
2. Sit in the front row when a seminar's title intrigues you
3. Irreproducible results can be blessings in disguise
4. Always have an audience for your experiments
5. Avoid boring people
6. Science is highly social
7. Leave a research field before it bores you

Chapter 6. From postdoctoral years at Cambridge University

1. Choose an objective apparently ahead of its time
2. Work on problems only when you feel tangible success may come in several years
3. Never be the brightest person in a room
4. Stay in close contact with your intellectual competitors
5. Work with a teammate who is your intellectual equal
6. Always have someone to save you

Chapter 7. From the first years on Harvard's faculty

1. Bring your research into your lectures
2. Challenge your students' abilities to move beyond facts
3. Have your students master subjects outside your expertise
4. Never let your students see themselves as research assistants
5. Hire spunky lab helpers
6. Academic institutions do not easily change themselves

Chapter 8. From the security of being recently tenured

1. Teaching can make your mind move on to big problems
2. Lectures should not be unidimensionally serious
3. Give your students the straight dope
4. Encourage undergraduate research experience
5. Focus departmental seminars on new science
6. Join the editorial board of a new journal
7. Immediately write up big discoveries
8. Travel makes your science stronger

Chapter 9. From working for PSAC

1. Exaggerations do not void basic truths
2. The military is interested in what scientists know, not what they think
3. Don't back schemes that demand miracles
4. Controversial recommendations require political backing

Chapter 10. From being ennobled in Stockholm

1. Buy, don't rent, a suit of tails
2. Don't sign petitions that want your celebrity
3. Make the most of the year following announcement of your prize
4. Don't anticipate a flirtatious Santa Lucia girl
5. Expect to put on weight after Stockholm
6. Avoid gatherings of more than two Nobel Prize winners
7. Spend your prize money on a home

Chapter 11. From bad decisions made in Harvard Yard

1. Success should command a premium
2. Channel rage through intermediaries
3. Be prepared to resign over inadequate space
4. Have friends close to those who rule
5. Never offer tenure to practitioners of dying disciplines
6. Become the chairman
7. Ask the dean only for what he can give

Chapter 12. From being edited by Harvard University Press

1. Be the first to tell a good story
2. A wise editor matters more than a big advance
3. Find an agent whose advice you will follow
4. Use snappy sentences to open your chapters
5. Don't use autobiography to justify past actions or motivations
6. Avoid imprecise modifiers
7. Always remember your intended reader
8. Read out loud your written words

Chapter 13. From watching top science emerge in the Biological Labs

1. Two obsessions are one too many
2. Don't take up golf
3. Races within the same building bring on heartburn
4. Close competitors should publish simultaneously
5. Share valuable research tools

Chapter 14. From managing cancer research

1. Accept leadership challenges before your academic career peaks
2. Run a benevolent dictatorship
3. Manage your scientists like a baseball team
4. Don't make midseason trades
5. Only ask for advice that you will later accept
6. Use your endowment to support science, not for long-term salary support
7. Promote key scientists faster than they expect
8. Schedule as few appointments as possible
9. Don't be shy about showing displeasure
10. Walk the grounds

Chapter 15. From heading the Cold Spring Harbor Laboratory

1. Avoid boring people
2. Delegate as much authority as possible
3. Institutions are either moving forward or they are moving backward
4. Always buy adjacent property that comes up for sale

5. Attractive buildings project institutional strength
6. Have wealthy neighbors
7. Be a friend to your trustees
8. All take and no give will disenchant your benefactors
9. Never appear upset when other people deny you their money
10. Avoid being photographed
11. Never dye your hair or use collagen
12. Make necessary decisions before you have to

James D. Watson was director of Cold Spring Harbor Laboratory in New York from 1968 to 1993 and is now its chancellor. He was the first director of the National Center for Human Genome Research of the National Institutes of Health, from 1989 to 1992. A member of the National Academy of Sciences and the Royal Society, he has received the Presidential Medal of Freedom, the National Medal of Science, and, with Francis Crick and Maurice Wilkins, the 1962 Nobel Prize in Physiology or Medicine.

A NOTE ON THE TYPE

The text of this book was set in Minion, a typeface produced by the Adobe Corporation specifically for the Mac-
intosh personal computer, and released in 1990. Designed
by Robert Slimbach, Minion combines the classic charac-
teristics of old-style faces with the full complement of
weights required for modern typesetting.

Composed by North Market Street Graphics,
Lancaster, Pennsylvania

Printed and bound by Berryville Graphics,
Berryville, Virginia

Designed by Anthea Lingeman